Springer Theses

Recognizing Outstanding Ph.D. Research

Aims and Scope

The series "Springer Theses" brings together a selection of the very best Ph.D. theses from around the world and across the physical sciences. Nominated and endorsed by two recognized specialists, each published volume has been selected for its scientific excellence and the high impact of its contents for the pertinent field of research. For greater accessibility to non-specialists, the published versions include an extended introduction, as well as a foreword by the student's supervisor explaining the special relevance of the work for the field. As a whole, the series will provide a valuable resource both for newcomers to the research fields described, and for other scientists seeking detailed background information on special questions. Finally, it provides an accredited documentation of the valuable contributions made by today's younger generation of scientists.

Theses are accepted into the series by invited nomination only and must fulfill all of the following criteria

- They must be written in good English.
- The topic should fall within the confines of Chemistry, Physics, Earth Sciences, Engineering and related interdisciplinary fields such as Materials, Nanoscience, Chemical Engineering, Complex Systems and Biophysics.
- The work reported in the thesis must represent a significant scientific advance.
- If the thesis includes previously published material, permission to reproduce this must be gained from the respective copyright holder.
- They must have been examined and passed during the 12 months prior to nomination.
- Each thesis should include a foreword by the supervisor outlining the significance of its content.
- The theses should have a clearly defined structure including an introduction accessible to scientists not expert in that particular field.

More information about this series at http://www.springer.com/series/8790

Thérèse Cantwell

Low Frequency Radio Observations of Galaxy Clusters and Groups

Doctoral Thesis accepted by
the University of Manchester, Manchester, UK

Author
Dr. Thérèse Cantwell
Jodrell Bank Centre for Astrophysics, School of Physics and Astronomy
The University of Manchester
Manchester, UK

Supervisor
Prof. Anna Scaife
Jodrell Bank Centre for Astrophysics, School of Physics and Astronomy
The University of Manchester
Manchester, UK

ISSN 2190-5053 ISSN 2190-5061 (electronic)
Springer Theses
ISBN 978-3-030-07430-2 ISBN 978-3-319-97976-2 (eBook)
https://doi.org/10.1007/978-3-319-97976-2

This Springer imprint is published by the registered company Springer Nature Switzerland AG
The registered company address is: Gewerbestrasse 11, 6330 Cham, Switzerland

It was of course an impossible task. But he was used to them. Dragging a rat all the way from the wood to the hole had been an impossible task. But it wasn't impossible to drag it a little way, so you did that, and then you had a rest, and then you dragged it a little way again... The way to deal with an impossible task was to chop it down into a number of merely very difficult tasks, and break each one of them *into a group of horrible hard tasks, and each one of* them *into tricky jobs, and each one of them...*

—Terry Pratchett, *The Bromeliad*

For my sister. I love you.

Supervisor's Foreword

Radio astronomy offers a unique perspective in astrophysics, revealing myriad phenomena across a vast range of scales and energies that are otherwise invisible to us. From the high-energy Universe, typified by spectacular active galactic nuclei, to the cold relic radiation of the Big Bang, the radio spectrum is full of information about how the Universe started and how it evolved and grew. Moreover, radio observations give us a window into a cosmic laboratory, where conditions exist that are simply impossible to replicate on Earth. These data augment our understanding of physics and allow us to progress scientifically beyond the constraints of terrestrial laboratories. Stretching over four decades in frequency, the radio spectrum encompasses a wide variety of astrophysical radiation processes. However, it is the low frequency (long wavelength) end of this spectrum that has historically been most difficult to explore. With radio wavelengths extending to tens of metres, this difficulty has been in part due to the structural problems in building receivers large enough to achieve resolutions anywhere near those needed for detailed astrophysical analysis; the corrupting behaviour of the ionosphere at low frequencies, coupled with the brightness of the Galactic background, and the crippling computing requirements to account for all of the other calibration and imaging effects make low frequency radio analysis one of the most demanding areas of astronomy research. As challenging as it may be, low frequency radio astronomy is also one of the most scientifically valuable probes of the Universe. As well as providing access to the redshifted neutral hydrogen line from the epoch of reionisation, it also probes a larger population of relativistic electrons that is accessible at higher radio frequencies. These synchrotron emitting electrons show us the true extent of galaxies, they reveal the large-scale structure of galaxy clusters and filaments, and they illuminate the structure of magnetic fields in the lowest density regions of the Universe. Magnetic fields are an ubiquitous ingredient of astrophysical structure, but the cosmic origin of these fields is still an open question. In this thesis, Dr. Cantwell uses low frequency radio measurements from the LOw Frequency ARray (LOFAR) and the Giant Metrewave Radio Telescope (GMRT) to examine magnetised plasmas in some of the largest astrophysical structures known: giant radio galaxies (GRGs) and radio haloes. Due to the observational limitations associated with capturing such huge astrophysical structures, giant radio galaxies are historically a poorly sampled population of objects; however, their preferential placement in the more rarified regions of the cosmic web makes them a uniquely important probe of large-scale structure. In particular, the polarisation of the radio emission from giant radio galaxies is one of the few tools available to us that can be used to measure magnetic fields in regions where the strength of those fields is a key differentiator for competing models of the origin of cosmic magnetism. This thesis presents new data on the giant radio galaxy NGC 6251. Polarisation analysis of these data reveals that the magnetic field strength in the locality of this giant radio galaxy is an order of magnitude lower than in other comparable systems. Such low frequency polarisation data are crucial for detailed analyses of magnetic structure, but they are also the most challenging observational data to work with. This thesis presents a beautifully coupled description of the technical and scientific analysis required to extract valuable information from such data, and as the new generation of low frequency radio telescopes reveals the larger population of giant radio galaxies, it will be a significant resource for future analyses.

Manchester, UK
July 2018

Prof. Anna Scaife

Preface

The detection of Mpc-scale emission, such as radio halos and radio relics, in galaxy clusters provides evidence that cosmic ray electrons, as well as cluster-scale magnetic fields are present in clusters. As such, radio observations of clusters provide a unique opportunity to study the non-thermal populations of the intra-cluster medium. The process responsible for this large-scale diffuse emission is still not fully understood. The current dominant model links the formation of radio halos with cluster mergers. However, research into the formation mechanism is limited by the relatively small number of known halos. Currently, there are of order 100 known halos compared with >100, 000 clusters. This thesis aims to increase the number of known halos by taking advantage of an optical parameter, the relaxation parameter, which links a cluster's optical properties with its dynamical state. If the production of a radio halo is linked to dynamical state, then the relaxation parameter could potentially be used to select clusters which host radio halos.

Observations of Faraday rotation in sources embedded in cluster or group environments offer an alternative method for probing cluster and group magnetic fields. In particular, the variance in the Faraday depth of embedded sources has previously been used to determine the magnetic field in a number of clusters. Determining the magnetic field in galaxy groups using the same method is more difficult due to the lower density, and therefore smaller Faraday depths, present in these environments. The Faraday depths in galaxy groups are expected to be of order $1-10$ rad m^{-2}. The LOw Frequency ARray (LOFAR) has the highest precision available with a Faraday depth resolution of $\sim$1 rad m^{-2}. At LOFAR frequencies, most extra galactic sources are expected to be depolarised. The ideal targets for polarisation studies with LOFAR are nearby giant radio galaxies with high degrees of polarisation. These sources are often found in low-density group environments, the precise environments we would like to study, which minimises the effect of depolarisation. This thesis aims to test LOFAR's polarisation capabilities using observations of the giant radio galaxy NGC 6251 which is located in a poor group environment.

In Chap. 1 of this thesis, I review the current understanding of galaxy clusters, groups, and radio galaxies. I also describe some of the astrophysical processes important to this thesis.

In Chap. 2, I discuss the interferometry and the process of calibrating interferometric data. I also describe some of the techniques used later in the thesis such as QU-fitting and RM synthesis.

In Chap. 3, I present my observations of the massive merging galaxy cluster MACSJ2243.3-0935. I report the discovery of a radio halo in MACSJ2243.3-0935, as well as a new radio relic candidate, using the Giant Metrewave Radio Telescope and the KAT-7 telescope. The radio halo is coincident with the cluster X-ray emission and has a largest linear scale of approximately 0.9 Mpc. I measure a flux density of 10.0 ± 2.0 mJy at 610 MHz for the radio halo. I discuss equipartition estimates of the cluster magnetic field and constrain the value to be of the order of 1 μG. The relic candidate is detected at the cluster virial radius where a filament meets the cluster. The relic candidate has a flux density of 5.2 ± 0.8 mJy at 610 MHz. I discuss possible origins of the relic candidate emission and conclude that the candidate is consistent with an infall relic.

In Chap. 4, I present my GMRT observations at 610 MHz of three disturbed galaxy clusters, A07, A1235 and A2055. No diffuse emission was observed in any of the three clusters. In order to place upper limits on the radio halo power in these clusters, I have injected simulated halos at different radio powers into the UVdata. A07 has a radio halo upper limit of $P_{610\mathrm{MHz}} = 1.5 \times 10^{24}$ W Hz^{-1}. A2055 has a radio halo upper limit of $P_{610\mathrm{MHz}} = 1.8 \times 10^{24}$ W Hz^{-1}. A1235 has a radio halo upper limit of $P_{610\mathrm{MHz}} = 5.8 \times 10^{23}$ W Hz^{-1}. These limits are below the $P_{610} - L_X$ relation and rule out bright radio halo in these clusters. I have identified these clusters as potential hosts for Ultra Steep Spectrum Radio Halo (USSRH). Observations with LOFAR should be capable of confirming whether or not these clusters host USSRH.

In Chap. 5, I present observations of the giant radio galaxy NGC 6251 with LOFAR HBA. NGC 6251 is a giant radio galaxy with a borderline FRI/FRII morphology located in a poor group. The images presented in this chapter are the highest sensitivity and resolution images of NGC 6251 at these frequencies to date. Analysis of the low frequencies spectral index did not reveal any change in the low frequency spectra when compared with the higher frequency spectral index. NGC 6251 is found to be either at equilibrium or slightly electron dominated, similar to FRII sources. I calculated the ages of the low-surface brightness extension of the northern lobe and the backflow of the southern lobe, which are only clearly visible at these low frequencies, to be 205 Myr $< t <$ 368 Myr and 209 Myr $< t <$307 Myr, respectively. This could indicate that these components are relics of an earlier epoch of activity.

I present the first detection of polarisation at 150 MHz in NGC 6251, including a weak detection of polarisation in the diffuse emission of the northern lobe. Taking advantage of the high Faraday resolution of LOFAR, I detect Faraday complexity in the knot of NGC 6251 and interpret the weaker component as emission from the lobe located behind the knot. I place an upper limit on the variance in the Faraday depth in the knot of NGC 6251 of $\sigma^2_{\mathrm{RM}} < 5 \times 10^{-3}$ rad^2 m^{-4} and an upper limit on the magnetic field in the group of $B < 0.2$ μG.

Manchester, UK
September 2017

Dr. Thérèse Cantwell

Declaration

No portion of the work referred to in this has been submitted in support of an application for another degree or qualification of this or any other university or other institute of learning.

Acknowledgements

The last 4 years would not have been possible without the help and support of many people.

Thank you to my family, my Mom Rioghnach, my Dad George, and my sister Georgina for always being there for me and for listening to the very many boring rants about data reduction problems.

Thank you to my two supervisors, Anna and Judith, first for giving me the opportunity to carry out the research in this thesis and secondly for guiding and encouraging me through the 4 years it took to complete.

Thanks to Deirdre, Luke, Poppy and Matt for providing an escape when writing was tough.

To all the people in our research group: Poppy (you get two mentions!), Chris R., Alex, Fernando, David, Justin, Hayden, Jeff, Minnie and Chris S.: Thanks for all the help and friendship.

Thank you to Denise Gabuzda for introducing me to radio astronomy and giving me the opportunity to experience research during my undergrad. It is in no small part due to her guidance and teaching that I began a Ph.D. in radio astronomy.

Finally thank you to Mark. Thank you for always being there for me. Thank you for cheering me up when I am down and encouraging me when I feel overwhelmed. Thank you for making sure I remembered to do things like sleep and eat while writing this thesis. This would not have been possible without you.

Contents

List of Figures

List of Tables

Chapter 1
Introduction

1.1 Galaxy Clusters and Groups

Galaxy clusters are the largest virialised structures in the Universe with typical masses of order $10^{15} M_{\odot}$. Most of this mass is composed of dark matter. The other 10–20% is contained in baryonic matter, with the mass in the hot Intra-cluster medium (ICM) being about 10 times larger than the mass contained in galaxies (Kravtsov and Borgani 2012; Brunetti and Jones 2014). The ICM was first detected in the X-ray band, emitting via thermal Bremsstrahlung, indicating that the ICM is a thermalised plasma (Voit 2005). However the detection of Mpc scale diffuse emission in the radio band provides evidence that cosmic ray electrons (CRe) are also present in the ICM, as are cluster-scale magnetic fields (Brunetti and Jones 2014; Feretti et al. 2012). Such emission was first discovered in the Coma cluster (Willson 1970) but has since been discovered in many other galaxy clusters (Feretti et al. 2012). As such, radio observations of clusters provide a unique opportunity to study the non-thermal populations of the ICM.

1.1.1 Cluster Dynamical State

X-ray observations of galaxy clusters show that there is a bimodality in the cluster population with some clusters hosting a dense core (cool core clusters) and others showing no evidence of such a core (non cool core clusters) (Hudson et al. 2010; Sanderson et al. 2006, 2009b). The temperature, density and entropy profiles of cool core (CC) clusters and non cool core (NCC) clusters show marked differences. CC clusters have systematically higher central densities compared with NCC clusters while NCC cluster have higher central entropies than CC clusters (Sanderson et al. 2009b; Cavagnolo et al. 2009). The cooling time of the central cores in CC clusters is less than the Hubble time (Hudson et al. 2010). The cores of these clusters were expected to lose most of their thermal energy in 10–100 Myr leading to the forma-

T. Cantwell, *Low Frequency Radio Observations of Galaxy Clusters and Groups*, Springer Theses, https://doi.org/10.1007/978-3-319-97976-2_1

tion of cooling flows where the peripheral gas moves subsonically towards the core (Gaspari et al. 2013; Fabian 1994). However the predicted cooling rates of order 1000s $M_{\odot}$ yr^{-1} were not observed (Peterson and Fabian 2006). Some additional mechanism is required to quench the expected cooling flows. It is currently thought that AGN feedback is responsible for the quenching of cooling flows in CC clusters (Croton 2006; Bower et al. 2006; Sijacki et al. 2007; Fabian 2012).

CC clusters are considered to be fully relaxed systems while NCC clusters are disturbed clusters (Sanderson et al. 2009a; Böhringer et al. 2010). There are a number of methods available to classify a cluster as either relaxed or disturbed (Mann and Ebeling 2012; Buote and Tsai 1995; Mohr et al. 1995; Hudson et al. 2010). One method is to compare the position offset between the peak in the ICM X-ray emission and the brightest cluster galaxy (BCG) (Sanderson et al. 2009a). During a cluster merger the galaxies in the merging clusters can be considered to be collisionless particles where as the ICM is a highly collisional gas. As the two dark matter potentials cross each other the member galaxies follow the dark matter wells while the ICM collides and no longer traces the dark matter potential. Thus a separation between the BCG and peak in the X-ray is expected for a merging system.

A new method of determining the dynamical state was introduced by Wen and Han (2013). They calculate the relaxation parameter, Γ, from the observed substructure in the optical luminosity of clusters. Positive values of Γ indicate a relaxed system and negative values of Γ indicate a disturbed cluster. The relaxation parameter separates known relaxed and unrelaxed clusters with a success rate of 94%.

Wen and Han (2013) define the relaxation parameter as

$$\Gamma = \beta - 1.90\alpha + 3.58\delta + 0.10, \tag{1.1}$$

where α is the asymmetry factor, β s the ridge flatness and δ is the normalised deviation. In order to calculate α, β and δ the optical image is first smoothed. The total fluctuation power in calculated as

$$S^2 = \sum_{\mathrm{i,j}} I^2(x_\mathrm{i}, y_\mathrm{j}), \tag{1.2}$$

where $I(x_\mathrm{i}, y_\mathrm{j})$ is the luminosity within pixel x_i, y_j. The difference power is given by

$$\Delta = \sum_{\mathrm{i,j}} \frac{\left[I(x_\mathrm{i}, y_\mathrm{j}) - I(-x_\mathrm{i}, -y_\mathrm{j})\right]^2}{2}, \tag{1.3}$$

so that

$$\alpha = \frac{\Delta^2}{S^2}. \tag{1.4}$$

A fully relaxed cluster would be expected to be very symmetric and would have $\alpha = 0$. Disturbed clusters are expected to be very asymmetric and should have an asymmetry factor closer to $\alpha = 1$.

In order to calculate the ridge flatness a one dimension King model (King 1962) must first be fit for various directions in the cluster. The 1-D King model is given by

$$I_{1-\mathrm{D}}(r) = \frac{I_0}{1 + \left(\frac{r}{r_0}\right)^2}, \tag{1.5}$$

where r_0 is the characteristic radius of the King model, r is the distance from the centre of the cluster in a given direction and I_0 is the luminosity at $r = 0$. Wen and Han (2013) define a steepness parameter $c_{200} = \frac{r_{200}}{r_0}$ where r_{200} is the radius within which the average density of the cluster is 200 times the critical density of the Universe. For a fully relaxed cluster c_{200} should be the same in all directions. The ridge flatness is defined as

$$\beta = \frac{c_{200,\mathrm{min}}}{\langle c_{200} \rangle}. \tag{1.6}$$

δ is first calculated by first fitting a 2-D elliptical King model to the smooth optical map,

$$I_{2\mathrm{D}}(x, y) = \frac{I_0}{1 + \left(\frac{r_{\mathrm{iso}}}{r_0}\right)^2}, \tag{1.7}$$

where I_0 and r_0 are defined as before and r_{iso} is given by

$$r_{\mathrm{iso}} = [x\cos(\theta) + y\sin(\theta)]^2 + \varepsilon^2 [x\sin(\theta) + y\cos(\theta)]^2, \tag{1.8}$$

where θ is the position angle of the ellipse and ε is the ratio of the semiminor to semimajor axes. The normalised deviation is defined in Wen and Han (2013) as

$$\delta = \frac{\sum_{\mathrm{i,j}} \left[I(x_{\mathrm{i}}, y_{\mathrm{j}}) - I_{2\mathrm{D}}(x_{\mathrm{i}}, y_{\mathrm{j}})\right]^2}{S^2}. \tag{1.9}$$

When clusters of known dynamical state are plotted in the 3D space of α, β and δ, the plane that separates known relaxed and unrelaxed clusters is described by Γ as given in Eq. 1.1.

1.1.2 Diffuse Radio Emission in Galaxy Clusters

The dynamics and evolution of galaxy clusters can also be indirectly probed using radio observations. Diffuse radio emission in clusters is divided into three morphological classes: radio relics; giant radio halos; mini halos. Both giant radio halos and radio relics have typical physical sizes of 1 Mpc. Due to the slow diffusion rate of CRe in clusters, coherent synchrotron emission on these scales requires in situ injection of CRe or re-acceleration of pre-existing, lower energy CRe (Brunetti and Jones 2014).

1.1.2.1 Radio Relics

Figure 1.1 shows a few examples of radio relics. Radio relics are normally elongated structures found at the periphery of clusters and can be highly polarised. Radio relics are associated with disturbed clusters, although not all merging systems host radio relics.

Radio relics are thought to trace weak shocks driven by major and minor cluster mergers (Brunetti and Jones 2014; Miniati et al. 2001; Brüggen et al. 2012; Kang et al. 2012). Traditionally, particle acceleration at these shocks is described by diffusive shock acceleration (DSA) (Jones and Ellison 1991; Kang et al. 2012). In DSA, particles are confined in a converging flow across the shock. Whenever the particle is reflected upstream across the shock it gains energy. In this picture, the magnetic field of the relic would be expected to align with the long axis of the shock. Older CRe would be found downstream of the shock and so a spectral index map would be expected to show a gradient across the relic, with the steepest spectral indices found at the injection point. Multi Frequency observations of radio relics provide support for this model (van Weeren et al. 2016). However recent work (Vazza and Brüggen 2014; Vazza et al. 2016) suggests that, given our current information on relic Mach numbers, cluster magnetic fields, and gamma ray flux upper limits, DSA requires either unrealistically large magnetic fields or predicts gamma ray fluxes above the upper limits. One proposed solution is that the shock front is reaccelerating a fossil population of low energy CRe, possible provided by the fading lobes of a radio galaxy (Markevitch et al. 2005). Observations of radio galaxies connected to radio relics appear to support this scenario (van Weeren et al. 2017; Bonafede et al. 2014).

1.1.2.2 Radio Halos

Giant radio halos are usually found at the center of clusters and typically have a more rounded morphology than radio relics. Giant radio halos tend to be largely unpolarised, likely due to some combination of the variation in the intrinsic polariation angle of the halo and the variation in the Faraday depth on scales smaller than the beam. Figure 1.2 shows two examples of radio halos.

One theory for the production of a giant radio halo is the hadronic model. In this model, collisions between CRp's and thermal protons produce $\pi^{\pm}$ which decay into CRe. Gamma rays are produced as a by-product of this reaction (Blasi and Colafrancesco 1999; Miniati et al. 2001; Enßlin et al. 2011). However, to date, there have been no gamma ray detections from the ICM, putting limits on the generation of secondary CRe that disfavour the hadronic model (Brunetti 2009; Jeltema and Pro-

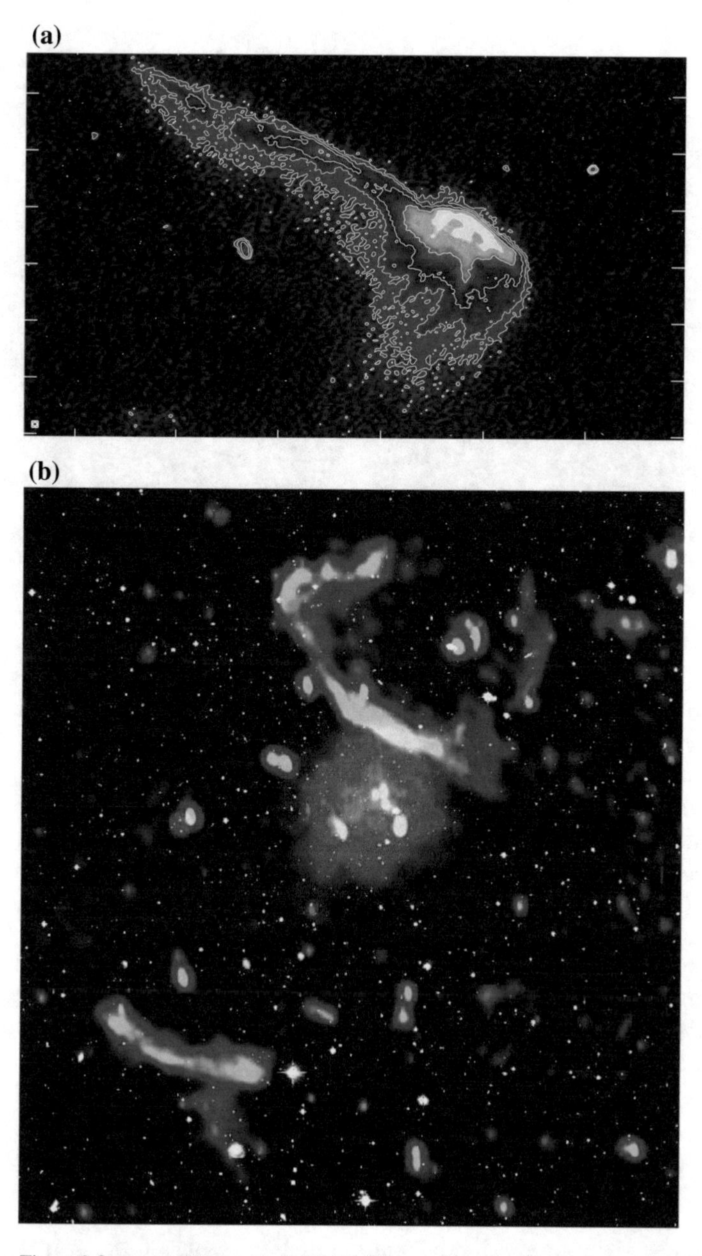

Fig. 1.1 **a** Figure 2 from van Weeren et al. (2016). The toothbrush relic as observed by LOFAR at 150 MHz. **b** Figure 1 from Bonafede et al. (2014). A double relic in the cluster PLCKG287.0 +32.9. X-ray observations with XMM Newton are shown in red while radio observations at 323 MHz are shown in blue

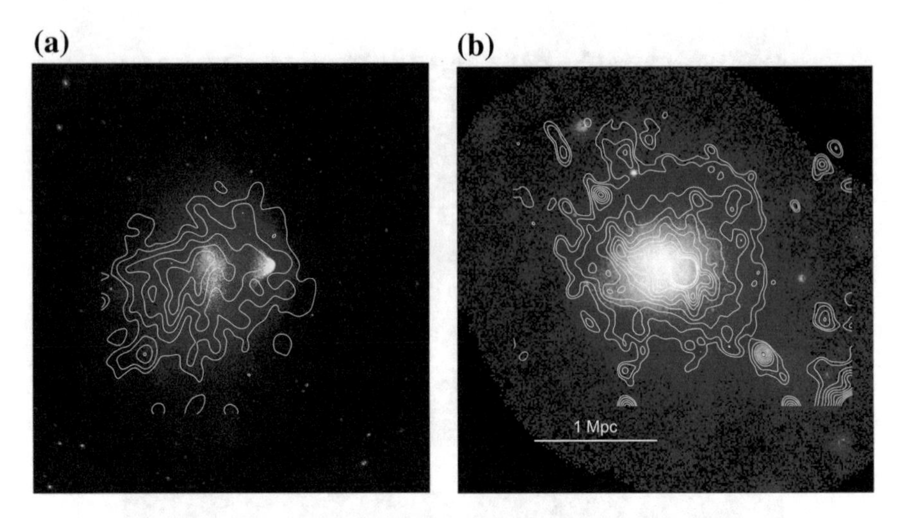

Fig. 1.2 **a** Figure 12a from Markevitch (2012) showing the contours of the radio halo in the bullet cluster overlaid on the cluster X-ray emission. **b** Figure 12f from Markevitch (2012) showing the contours of the radio halo in the Coma cluster overlaid on the cluster X-ray emission

fumo 2011; Brunetti et al. 2012).The broad surface brightness profiles of some radio halos are also difficult to reproduce with the hadronic model as this would require a radially increasing magnetic field or energy density of CRp (Donnert et al. 2010; Brunetti and Jones 2014; Brunetti 2003, 2004; Marchegiani et al. 2007; Brunetti and Jones 2014; Zandanel et al. 2014).

An alternative model to the hadronic model is the reacceleration model, where a fossil population of low energy CRe is reaccelerated to high energies by turbulence in the ICM (Petrosian 2001; Brunetti et al. 2001; Fujita et al. 2003; Cassano and Brunetti 2005; Donnert et al. 2013). Major mergers are thought to be the driving force for this turbulence, as radio halos are found almost exclusively in clusters undergoing such mergers (Donnert et al. 2013; Brunetti et al. 2009). The discovery of a number of radio halos in CC clusters posses a challenge to this model (Venturi et al. 2017; Sommer et al. 2017; Bonafede et al. 2014).

1.1.3 Faraday Rotation in Galaxy Clusters

The presence of diffuse synchrotron radio emission associated with the ICM is proof of the existence of large scale magnetic fields in galaxy clusters. Faraday Rotation (see Sect. 1.3.3) offers a method to probe the magnetic field of the ICM. The magnetised thermal gas of the ICM will rotate the polarised signal of background (or embedded) radio galaxies. Following Govoni and Feretti (2004), if it is assumed that the turbulent ICM is composed of cells with a uniform density, uniform magnetic field, random field orientation and a characteristic scale length Λ_c then the average Faraday depth measured for these sources is $\langle RM \rangle = 0$ and the variance is

$$\sigma_{\rm RM}^2 = 812^2 \Lambda_c[kpc] \int \left(B_{\|}[\mu G] n_e \left[cm^{-3}\right]\right)^2 \ \text{rad}^2\,\text{m}^{-4}. \tag{1.10}$$

Thus measuring the variance in RM of sources embedded in a cluster can be used to estimate the magnetic field in a galaxy cluster. This has been done for a number of galaxy clusters (Govoni et al. 2010; Clarke et al. 2001; Bonafede et al. 2013; Feretti et al. 1999; Govoni et al. 2006). These surveys find that embedded radio galaxies have patchy Faraday depths structure of order several hundred rad m^{-2} on scales of 5–15 kpc. The magnetic field strengths measured are of the order of a few μG with σ_{rm} decreasing with increasing distance from the cluster centre.

1.1.4 Radio Halo Scaling Relationships

Since their discovery, a number of empirical scaling relations have been found between the radio power of giant radio halos and the properties of the host cluster such as cluster mass, temperature, and X-ray luminosity (Colafrancesco 1999; Govoni et al. 2001; Feretti 2002; Enßlin and Röttgering 2002; Feretti 2003; Brunetti et al. 2009; Cassano et al. 2013; Yuan et al. 2015). The most well studied scaling relationship is between the radio power at 1.4 GHz, $P_{1.4}$, and the X-ray luminosity, $L_{\rm X}$, of the ICM with more powerful halos found in high luminosity clusters. Figure 1.3 shows the $P_{1.4} - L_{\rm X}$ as presented in Cassano et al. (2013). There are two distinct populations seen in Fig. 1.3. The first is the population of radio halos, which exhibits the correlation between $P_{1.4}$ and $L_{\rm X}$. The second is the population of radio quiet clusters, marked as upper limits in Fig. 1.3. The upper limits on $P_{1.4}$ lie well below the expected $P_{1.4}$ values from the $P_{1.4} - L_{\rm X}$ correlation. Cassano et al. (2010) show that these two populations can be separated based on cluster dynamical state with dynamically disturbed clusters hosting halos and relaxed clusters devoid of diffuse radio emission. Brunetti et al. (2009) examine the bimodality in the $P_{1.4} - L_{\rm X}$ diagram and find that given the rarity of halos with powers well below the $P_{1.4} - L_{\rm X}$ correlation, the timescales for amplification and suppression of synchrotron emission in clusters must be short. They show that this is most easily explained by the reacceleration model where the high frequency synchrotron emission rapidly decreases post merger due to the dissipation of a significant amount of the generated turbulence.

1.1.5 Galaxy Groups

Galaxy groups are less massive counterparts to galaxy clusters, with masses of order $\sim 10^{14} M_{\odot}$. While they are less massive than galaxy clusters, galaxy groups are more

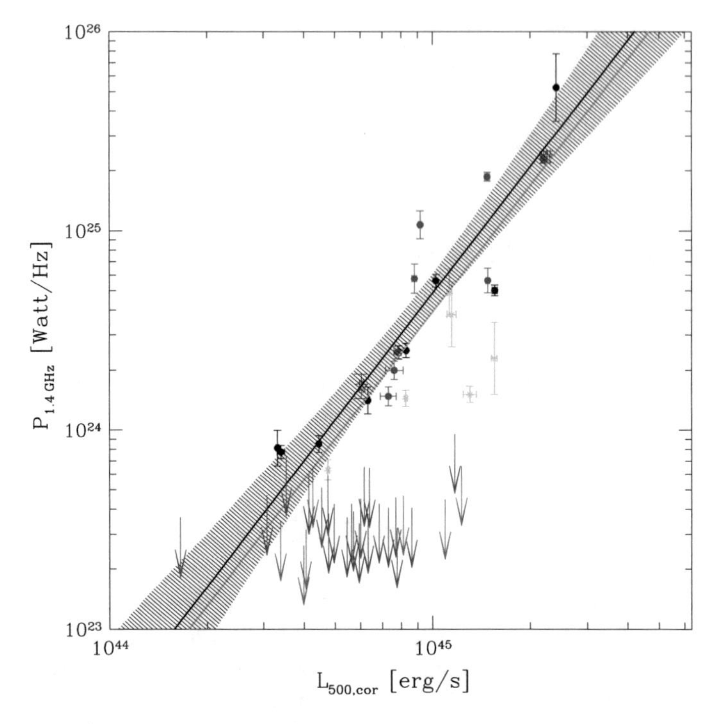

Fig. 1.3 Figure 2b from Cassano et al. (2013) showing the $P_{1.4} - L_X$ scaling relationship for radio halos

common and contain approximately 70% of the galaxies in the Universe (Tully 2015; Mulchaey 2000; Tully 1987). Galaxy groups are unique astrophysical structures and not simply scaled down versions of galaxy clusters, showing systematic differences in observed scaling relations such as the $L_X - M$ relationship (Ponman et al. 1999, 2003; Sun 2012; Paul et al. 2017). However significantly less work has been done in relation to galaxy groups compared with galaxy clusters.

Radio observations of poor groups could prove important in constraining origin models for cosmic magnetic fields. The magnetic fields observed in clusters can be replicated by many models and detection of magnetic fields in rarer environments, such as groups, is needed to properly constrain origin models (Vazza 2016; Donnert et al. 2009).

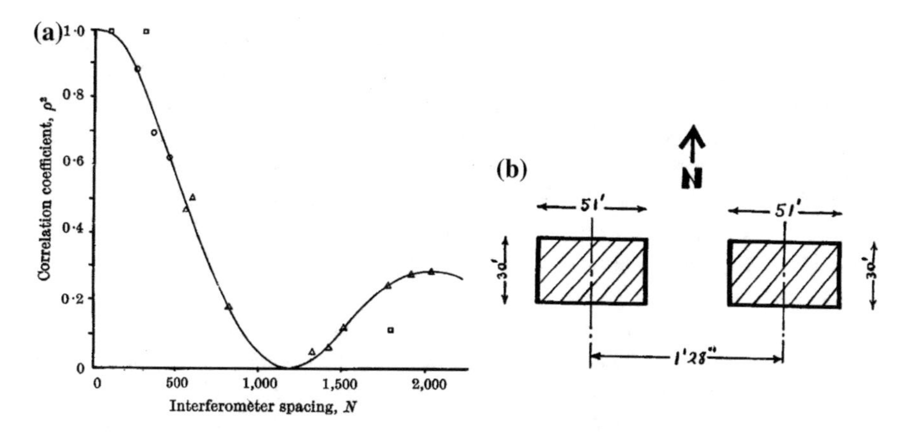

Fig. 1.4 **a** Intensity distribution of Cygnus A as measured by an interferometer at Jodrell Bank Observatory. Reproduced from Fig. 1 in Jennison and Das Gupta (1953). **b** Model of Cygnus A proposed by Jennison and Das Gupta (1953) using the intensity distribution shown in **a**. Reproduced from Fig. 2 in Jennison and Das Gupta (1953)

1.2 Active Galactic Nuclei

1.2.1 Radio Galaxies

The first radio source identified as having an extra galactic origin was Cygnus A. Baade and Minkowski (1954) identified two merging optical galaxies as the optical counter parts of Cygnus A, making it the first radio galaxy identified. Later Jennison and Das Gupta (1953) used a two element interferometer at Jodrell Bank Observatory to observe Cygnus A. Using the intensity distribution shown in Fig. 1.4, Jennison and Das Gupta (1953) demonstrated that the radio source was not a single component centred on the merging galaxies but two regions of emission of approximately 51 arcsec each separated by more than 1 arcmin. The total physical extent of the source is significantly larger than the physical size of the merging galaxies and showed that the radio emission was not from the merging galaxies. This model was confirmed by observations of Cygnus A with the one-mile radio telescope and later the 5 km radio telescope at Cambridge (Ryle et al. 1965; Hargrave and Ryle 1974).

At the same time that Jennison and Das Gupta (1953) showed that Cygnus A was a double structure, observations showed that the optical jet in M87, known since 1918, was polarised synchrotron emission (Baade 1956). The M87 jet was then assumed to be partially responsible for the radio emission associated with M87 (Baade and Minkowski 1954; Burbidge 1956). Increasingly higher resolution radio observations of M87 over the next few decades confirmed that there is a radio counterpart to the optical jet (Graham 1970; Turland 1975).

The construction of the 5 km radio telescope at Cambridge and later high resolution radio telescopes allowed a clear picture of the radio galaxy structure to be built.

Morphologically radio galaxies can largely be separated into two different categories, Fanaroff-Riley Class I (FRI) and Fanaroff-Riley Class II (FRII) (Fanaroff and Riley 1974). FRI sources are those for which the distance between the two brightest regions on either side of the source is less than half the total extent of the source. FRI sources typically have a main jet and a counter jet. The lobes of FRI sources are typically either plumed, with most of the lobe emission beyond the end point of the collimated jet, or bridged, where most of the lobe emission is between the end of the jet and the nucleus. Figure 1.5a shows 3C31, an example of a plumed FRI while Fig. 1.5b shows Hercules A, an example of a bridged FRI.

FRII sources are those for which the distance between the two brightest regions on either side of the source is more than half the total extent of the source. FRII sources typically have only one visible jet and in general this jet is well collimated and straight, in comparison with the puffy, wavy FRI jets. FRII jets terminate in very bright knots known as hotspots. A hotspot is visible in both lobes, even when only one jet is visible. The hotspots are embedded in the lobes of the FRII source. Figure 1.5c shows Cygnus A, an example of an FRII source.

FRII sources are much more powerful than FRI sources. Initially it was thought that these sources could be separated based on their radio power with FRII sources having a radio power above 10^{25} W Hz^{-1} at 1.4 GHz and FRI sources having a radio power below this. However when the radio luminosity at 1.4 GHz is plotted against the optical luminosity it is found that the break depends on the magnitude of the host galaxy (Owen 1993; Owen and Ledlow 1994). This suggests that environment has a role to play in the morphology of radio galaxies.

1.2.2 *Quasars, Blazars, and Seyferts*

A small percentage of galaxies, mainly spiral galaxies, are what is known as a Seyfert galaxy (Maiolino and Rieke 1995). Optically a Seyfert galaxy looks like a normal spiral galaxy but with a high surface brightness nucleus (Davidson and Netzer 1979). Seyfert galaxies can broadly be divided into two categories, Seyfert 1 and Seyfert 2 galaxies, based on the properties of their spectra (Khachikian and Weedman 1974). The spectra of Seyfert 1 galaxies show broad permitted lines, with velocities of order 10000 km s^{-2} as well as narrow forbidden lines with velocities of order 1000 km s^{-1} (Longair 2011). On the other hand the spectra of Seyfert 2 galaxies show that the forbidden lines and permitted lines have similar narrow widths, corresponding to velocities of order 1000 km s^{-1} (Longair 2011). The presence or absence of broad lines is the determining factor in the identification of a galaxy as Seyfert 1 or Seyfert 2. The interpretation is that the narrow and broad lines originate from different regions in the galaxy, with the narrow lines emitted further from the nucleus. The emission from Seyfert galaxies is variable on timescales of order several years (Peterson et al. 1984, 1998).

When radio interferometry made it possible to determine the precise location of radio galaxies, much work was done to find optical counterparts for these sources.

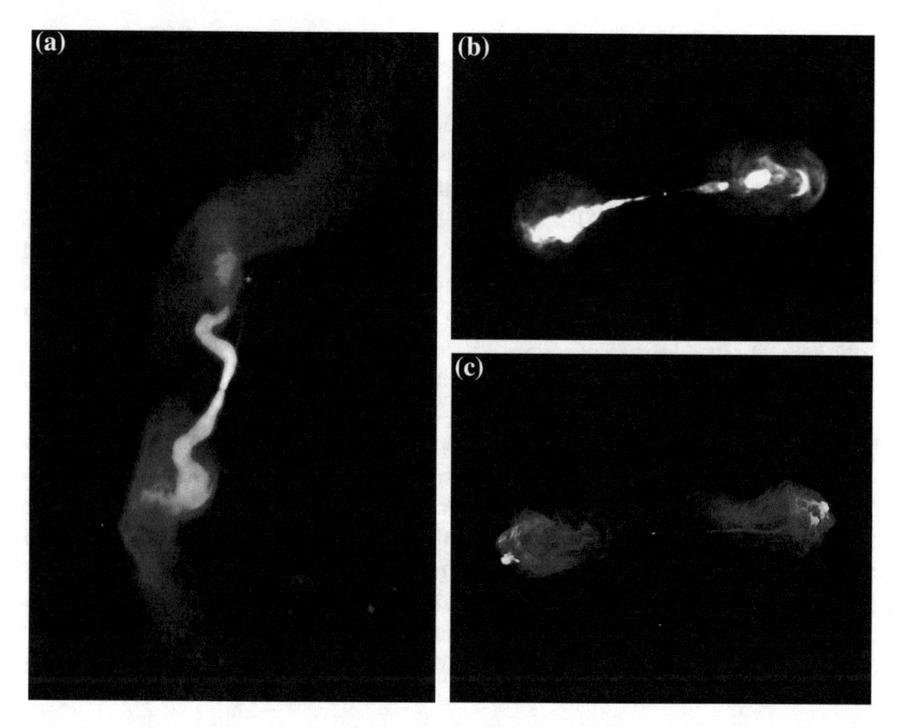

Fig. 1.5 **a** Plumed FRI 3C31 (Laing et al. 2008, Image courtesy of NRAO/AUI). **b** Bridged FRI Hercules A. (Gizani and Leahy 2003) **c** FRII Cygnus A (Carilli and Barthel 1996, Image courtesy of NRAO/AUI)

A number of radio galaxies were associated with unusual optical sources. These sources are stellar like but with optical spectra that could not be identified at the time (Matthews and Sandage 1963). Schmidt (1963) recognised the optical spectra of one of these sources, 3C273, as the Balmer lines of hydrogen, only redshifted to $z = 0.158$. These sources became known as quasi-stellar radio sources, or quasars. Quasars are the extreme end of the Seyfert population and are intrinsically more luminous with Seyfert 1 like spectra (Davidson and Netzer 1979). Quasars also vary on timescales of months to years (Hook et al. 1994; Vanden Berk et al. 2004; MacLeod et al. 2010).

Blazars are another extreme type of quasar. BL Lac blazars (named for the prototypical source BL lacerate) are highly variable radio sources with a high degree of polarisation (Padovani and Giommi 1995; Angel and Stockman 1980). These radio sources are only resolved at VLBI resolutions which show one sided jets with superluminal velocities suggesting that the jets must be aligned close to the line of sight (Angel and Stockman 1980; Zensus 1989). Unlike quasars, BL Lacertae sources do not have broad line regions. Optically violently variable (OVV) blazars are variable in both the radio and the optical and their spectra do show a broad line region (Longair 2011).

1.2.3 A Unified Model for AGN

A lot of effort has been made to unify the disparate AGN classes into a single source type where different AGNclasses are observed depending on properties such as the environment of the AGN and the inclination of the source to the line of sight (Netzer 2015; Heckman and Best 2014; Antonucci 1993; Elitzur 2012). Figure 1.6 shows an example of the unified model of AGN, reproduced from Fig. 3 in Heckman and Best (2014). The model predicts different behaviour, depending on the accretion mode. Here we will discuss the basic properties of the unified model as presented in Heckman and Best (2014).

In the radiative mode, the AGN is composed of a black hole which is accreting material via a radiatively efficient thin disk. Ionising radiation from the disk photoionises a population of dense clouds near the central black hole and is the region responsible for the broad line emission. The accretion disk is surrounded by a torus of dusty gas. The ionising radiation that escapes via the polar axis of this structure then ionises low density gas at kpc scales. This low density region is responsible for the narrow forbidden line emission. Some of the more massive radiative mode AGN can produce powerful radio jets. Different viewing angles lead to the different observed emission. Lines of sight through the obscuring structure block the broad line emission, leading to Seyfert 2 type galaxies, while lines of sight through the polar

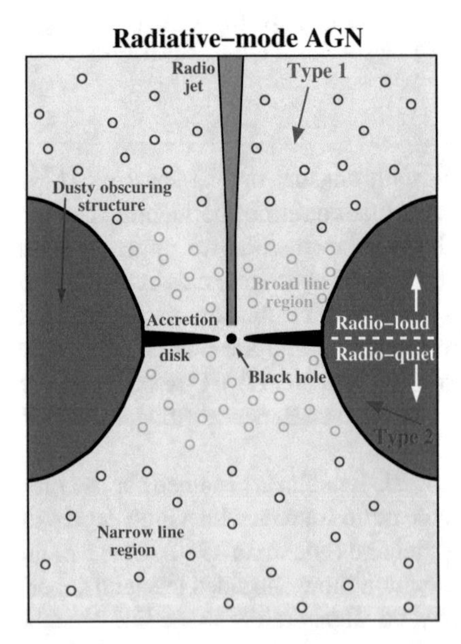

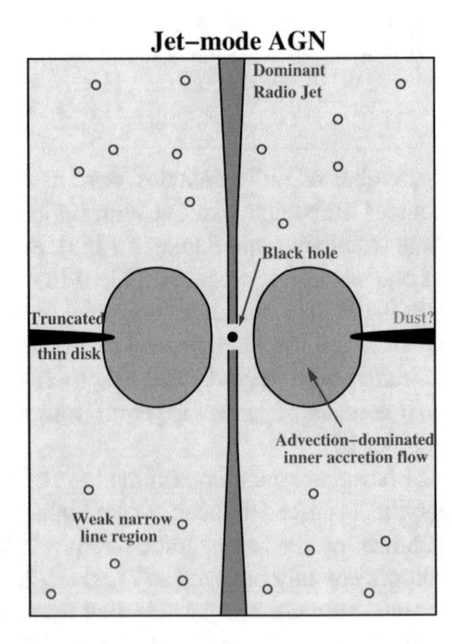

Fig. 1.6 **a** Schematic of a radiative-mode AGN. The top half of the image depicts the radio loud scenario and the bottom half of the image depicts the radio quiet scenario. The red arrows depict the viewing angle giving rise to type 1 and type 2 Seyfert galaxies. **b** Schematic of a jet mode AGN. Figure 3 in Heckman and Best (2014)

axis of the torus allow the detection of both regions and are identified as Seyfert 1 type galaxies.

In jet mode AGN, the accretion is low and radiatively inefficient. Instead of a thin accretion disk there is advection dominated accretion flow. Such accretion flows naturally produce twin jets. As there is no accretion disk the narrow line region is either absent or faint for these AGN.

1.2.4 AGN and Feedback

Feedback from AGN jets has been shown to play an important role in regulating the cooling of galaxies, groups, and clusters (Croton 2006; Bower et al. 2006; Sijacki et al. 2007; Fabian 2012). Here we will discuss some of the observational evidence for AGN feedback.

Observations of groups and clusters show that often the lobes of the central radio galaxy, if present, are coincident with depressions in the thermal X-ray emission. The implication is that the expanding lobes of the radio galaxy evacuate a cavity in the ICM. The energy required to create such a cavity can be calculated from the pressure and volume of the cavity as well as the energy in the lobe such that $E_{\mathrm{cav}} = E + pV$. This energy can vary from 2 to 4 times pV (Fabian 2012). If the age of the cavity is estimated it is possible to use E_{cav} as an estimate for the mechanical energy of the jet that is available to heat the cluster. In general this is only possible for low redshift clusters and groups, due to the difficulty of measuring the X-ray emission of the thermal gas at higher redshifts.

In order to study feedback from AGN at redshifts where detections of X-ray cavities are not possible, it is necessary to have some method of estimating the jet power. Attempts have been made to estimate the jet power using a correlation between jet power and radio luminosity, calibrated using X-ray cavities (Cavagnolo et al. 2010; Best et al. 2014; Kokotanekov et al. 2017). Unfortunately a large amount of scatter in these scaling relations prevent the application to higher redshift sources. One source of scatter is the uncertainty in the exact particle content in radio jets and lobes (Godfrey and Shabala 2016). As such it is critically important to understand the particle content in radio galaxies.

1.2.5 Particle Energetics

It is possible to model the particle content in radio jets by comparing the external pressure, as determined from X-ray observations, with the internal pressure estimated from radio observations. The most simple assumption one can make when calculating the internal pressure is that the relativistic electrons and magnetic field are in equipartition (Burbidge 1956; Hardcastle et al. 2002; Laing and Bridle 2002; Croston et al. 2004). Such pressure comparisons have been carried out for many sources. In general it is found that FRII sources are either at equipartition or slightly electron dominated with $\frac{U_e}{U_B} \sim 5$, where U_e is the electron energy density and U_B is the magnetic field energy density (Hardcastle and Worrall 2000; Croston et al. 2005; Shelton et al. 2011; Isobe and Koyama 2015; Kawakatu et al. 2016; Ineson et al. 2017). In contrast FRI sources are found to be underpressured when equipartition is assumed (Morganti et al. 1988; Worrall and Birkinshaw 2000; Croston et al. 2008; Croston and Hardcastle 2014).

Many models have been proposed to reconcile the internal and external pressures of FRI radio galaxies such as magnetic field dominance (Li et al. 2006; Nakamura et al. 2006) and relativistic proton dominance (De Young 2006; McNamara and Nulsen 2007; Bîrzan et al. 2008). The model most supported in the literature is the entrainment model, where thermal particles are entrained by the jet from its environment and these non-radiating particles provide the missing pressure (Hardcastle et al. 2003, 2007; Croston et al. 2008; Croston and Hardcastle 2014).

1.2.6 Giant Radio Galaxies

giant radio galaxies (GRG) are a population of radio galaxies with projected linear sizes greater than 1 Mpc. These sources are typically found in groups and are generally either FRII sources or borderline FRI/FRII (Ishwara-Chandra and Saikia 1999). Due to their large physical extent nearby GRG allow very detailed analysis of the jet and lobe structure of radio galaxies (Laing et al. 2006; Perley et al. 1984) as well as variations in the spectral index across the source (Mack et al. 1997, 1998).

The question of the origin of the huge sizes of GRG has been studied by many authors (Komberg and Pashchenko 2009; Subrahmanyan et al. 2008; Machalski et al. 2002; Saripalli et al. 1997; Mack et al. 1998). Multifrequency radio observations of GRG show that GRG are not significantly older than their kpc scale counterparts (Mack et al. 1997; Saripalli et al. 2005). Age is therefore not the cause of the larger projected size of GRG. Instead GRG are thought to grow to Mpc scales due to the low density of their host environment (Pirya et al. 2012; Machalski et al. 2011; Machalski and Jamrozy 2006). X-ray observations of the intergalactic medium (IGM) around GRG combined with spectroscopic observations of the group galaxies show that the X-ray luminosity of the IGM is much lower than would be expected from the $L_X - \sigma_r$ relationship (Chen et al. 2011, 2012).

The lobes of GRG extend beyond their host group to probe the large scale structure (LLS) of the Universe. Many GRG exhibit asymmetries in their source structure, suggesting interactions with the LLS (Pirya et al. 2012; Schoenmakers et al. 2000; Lara et al. 2001). Pirya et al. (2012) find that the shorter jet/lobe tend to point towards galaxy overdensities.

The lobes of GRG are potentially powerful indirect probes of the warm hot intergalactic medium (WHIM) that exists in large scale filaments. The WHIM is a natural prediction of λCDM cosmology and is thought to contain ~50% of the baryonic matter in the Universe (Davé et al. 2001; Nicastro et al. 2008; Smith et al. 2011).

To date there have been no direct detection of the WHIM due to the expected low density. Hence indirect measurements of the WHIM using observations of GRG is an important tool. This is done by assuming that the lobes of GRG are relaxed and in equilibrium with the external WHIM pressure. By calculating the internal pressure of the lobe we can therefore measure the pressure in the WHIM (Subrahmanyan et al. 2008; Safouris et al. 2009). Indeed Malarecki et al. (2015) combine radio observations of GRG with spectroscopic optical observations of nearby galaxies to demonstrate that it is possible to use GRG to probe the densest 6% of the WHIM.

Polarised observations of giant radio galaxies can probe the magnetic field of the host cluster or group. As mentioned in Sect. 1.1.3 there have been a number of studies using the rotation measures of radio galaxies to measure the magnetic fields in galaxy clusters. However much less work as been done for galaxy groups. Schoenmakers et al. (1998) use polarised observations of the GRG WNB 0313+683 to show that the mean magnetic field in the group is 0.5 μG. The low density environment of GRG is expected to have very low rotation measures, of order 1–10 rad m^{-2}, as well as small RMdispersions. Precise measurements of this value require high RM precision. LOFAR has the highest precision available with a Faraday depth resolution of $\sim$1 rad m^{-2}. Unfortunately at LOFAR frequencies, most extra galactic sources are expected to be depolarised. Mulcahy et al. (2014) presented the first detections of extragalactic polarisation with LOFAR and found approximately 1 source per 1.7 square degrees. Orrù et al. (2015) also report the detection of polarisation in the outer lobes of the double-double radio galaxy B1834+620. The ideal targets for polarisation studies with LOFAR are nearby giant radio galaxies with high degrees of polarisation. These sources are found in low density environments, the precise environments we would like to study, which minimises the effect of depolarisation.

1.3 Astrophysical Processes

1.3.1 Synchrotron Emission

Most extragalactic radio sources are synchrotron emitters. synchrotron radiation is emitted by relativistic charged particles moving in an electric field. These particles will move in a spiral about the field lines. A complete explanation of synchrotron radiation can be found in Longair (2011) and Rybicki and Lightman (1986). Here we reproduce some of the results for reference. To calculate the rate of energy loss for a single electron we consider the coordinate system shown in Fig. 1.7 which is in the instantaneous rest frame of the electron. Since the instantaneous velocity in the rest frame is $v' = 0$ we have

$$\gamma m \dot{\mathbf{v}}' = e\mathbf{E}' + \left(\mathbf{v}' \times \mathbf{B}'\right) = e\mathbf{E}'. \tag{1.11}$$

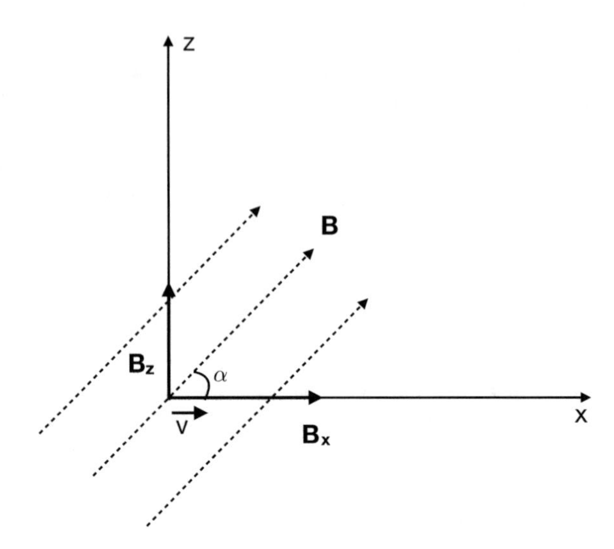

Fig. 1.7 Coordinate frame for relativistic electron moving in a magnetic field

The electric field in the instantaneous rest frame of the electron is given by

$$\begin{aligned} E'_{\mathrm{x}} &= E_{\mathrm{x}} = 0, \\ E'_{\mathrm{y}} &= \gamma \left(E_{\mathrm{y}} - vB_{\mathrm{z}}\right) = -\gamma v B_{\mathrm{z}}, \\ E'_{\mathrm{z}} &= \gamma \left(E_{\mathrm{z}} - vB_{\mathrm{y}}\right) = 0, \end{aligned} \tag{1.12}$$

so that

$$\dot{\mathbf{v}}' = -\frac{e\gamma v B_{\mathrm{z}}}{m_{\mathrm{e}}}. \tag{1.13}$$

The rate of energy loss is then given by the Larmor formula

$$\begin{aligned} -\frac{dE'}{dt'} &= \frac{e^2 \left|\dot{\mathbf{v}}'\right|}{6\pi\varepsilon_0 c^3}, \\ -\frac{dE'}{dt'} &= \frac{e^4\gamma^2 v^2 B^2 \sin^2\alpha}{6\pi\varepsilon_0 c^3 m_{\mathrm{e}}^2}, \\ -\frac{dE'}{dt'} &= 2\sigma_{\mathrm{T}} \left(\frac{v}{c}\right)^2 \frac{cB^2}{2\mu_0}\gamma^2 \sin^2\alpha, \end{aligned} \tag{1.14}$$

which is the same as the energy loss in the lab frame as $-\frac{dE}{dt}$ is a Lorentz invariant. Since $E = \gamma m_{\mathrm{e}} c$, $-\frac{dE}{dt} \propto B^2 E^2$. Therefore in a uniform mangetic field electrons with higher energies will lose their energy faster than lower energy electrons. Over time the effect of these synchrotron losses leads to steepening of the synchrotron spectra at higher frequencies.

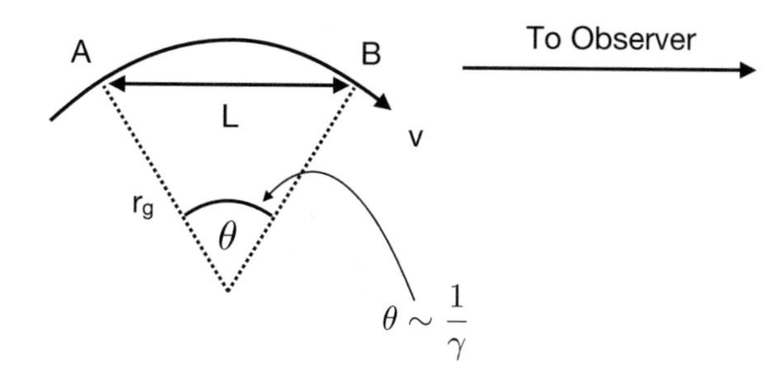

Fig. 1.8 Schematic of region of electrons orbit for which synchrotron emission is observed

Because synchrotron emitting electrons are relativistic the radiation is beamed in the direction of the electrons motion. From the Larmor formula the angular power spectrum is proportional to $\cos^2\theta'$ in the instantaneous restframe of the electron. The angular power spectrum has null points at $\theta' = \frac{\pi}{2}$. Converting to the lab frame using the abberation formula we have

$$\sin\theta = \frac{1}{\gamma}\left(\frac{\sin\frac{\pi}{2}}{1+\frac{v}{c}\cos\frac{\pi}{2}}\right) = \frac{1}{\gamma}. \tag{1.15}$$

For relativistic electrons $\frac{1}{\gamma} \ll 1$ and so $\theta \sim \frac{1}{\gamma}$. Therefore emission is only observed for $\frac{1}{\gamma}$ radians of the electrons orbit.

Figure 1.8 shows the diagram of this visible region. If the distance to the observer is R than and emission leaves point A at time $t_0 = 0$ then the radiation arrives at the observer at $t_1 = \frac{R}{c}$. Radiation leaves point B at $t_2 = \frac{L}{v}$ and arrives at the observer at time $t_3 = t_2 + \frac{R-L}{c}$. The observed duration of the pulse is thus

$$\begin{aligned}\Delta t &= \left[\frac{L}{v} + \frac{R-L}{c}\right] - \frac{R}{c} \\ &= \frac{L}{v}\left(1 - \frac{v}{c}\right).\end{aligned} \tag{1.16}$$

Since $\frac{L}{v} = \frac{r_g\theta}{v} \approx \frac{1}{\gamma\omega_r}$ and $\left(1 - \frac{v}{c}\right) \approx \frac{1}{2\gamma^2}$ we have

$$\Delta t \approx \frac{1}{2\gamma^3\omega_r}. \tag{1.17}$$

This corresponds to a frequency of $\nu \sim \Delta t^{-1}$ or

$$\nu \sim \gamma^3\omega_r. \tag{1.18}$$

A more complete analysis shows that the spectrum of synchrotron radiation for a single electron, shown in Fig. 1.9 is strongly peaked at a critical frequency which is equal to the frequency derived above (Longair 2011; Rybicki and Lightman 1986).

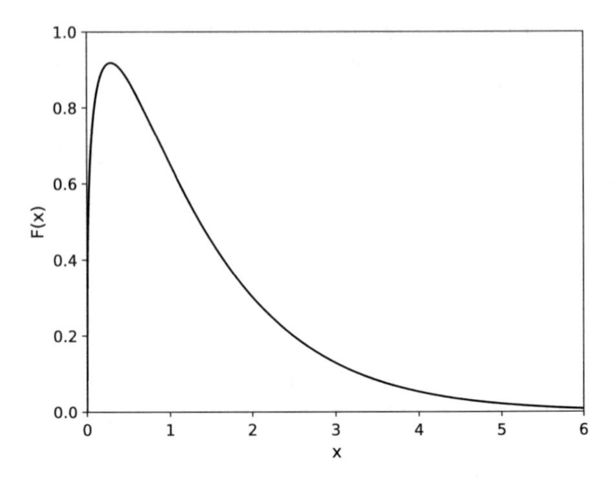

Fig. 1.9 Synchrotron spectrum for a single electron

We can therefore assume that most energy is radiated away at this frequency so that the emissivity for a population of electrons with distribution $N(E)dE = \kappa E^{-p} dE$ is given by

$$J(\nu)d\nu = -\left(\frac{dE}{dt}\right) N(E)dE. \tag{1.19}$$

Substituting in Eqs. 1.14 and 1.18 we find

$$J(\nu) \propto \kappa B^{\alpha+1} \nu^{-\alpha}, \tag{1.20}$$

where $\alpha = \frac{(p-1)}{2}$.

Synchrotron emission is expected to be polarised. In the case of non relativistic particles or weakly relativistic particles (cyclotron radiation), when the magnetic field is perpendicular to the line of sight the acceleration vector for the electron oscillates in the plane parallel to the line of sight leading to linearly polarised emission. When the magnetic field is parallel to the line of sight, the polarisation is circularly polarised while for any other orientation of the magnetic field to the line of sight will be elliptically polarised. The fractional polarisation is defined as

$$\Pi = \frac{I_\perp(\omega) - I_\parallel(\omega)}{I_\perp(\omega) + I_\parallel(\omega)} \tag{1.21}$$

where $I_\perp(\omega)$ is the component of the intensity perpendicular to magnetic field and $I_\parallel(\omega)$ is the component of the intensity parallel to the magnetic field. For a power law distribution of electron energies the fraction polarisation is

$$\Pi = \frac{p+1}{p+\frac{7}{3}}. \tag{1.22}$$

For an electron population with $p = 2.4$ the fraction is $\sim$72%.

1.3.2 Bremsstrahlung Radiation

Galaxy clusters are some of the most X-ray luminous sources in the Universe with typical luminosities in the range of $10^{43} - 10^{45}$ erg s^{-1} (Sarazin 1988). The emission mechanism responsible for the observed X-rays is Bremsstrahlung radiation from the hot intra cluster medium (Sarazin 1988; Felten et al. 1966). A complete explanation of Bremsstrahlung can be found in Longair (2011) and Rybicki and Lightman (1986). For reference we will reproduce some of the results here.

The acceleration of an electron moving at high velocity past a stationary nucleus in the electron rest frame is given by

$$\begin{aligned} \dot{v}_{\parallel}(t) &= \frac{\gamma Z e^2 v t}{4\pi\varepsilon_0 m_e \left[b^2 + (\gamma v t)^2\right]^{\frac{3}{2}}}, \\ \dot{v}_{\perp}(t) &= \frac{\gamma Z e^2 b}{4\pi\varepsilon_0 m_e \left[b^2 + (\gamma v t)^2\right]^{\frac{3}{2}}}, \end{aligned} \tag{1.23}$$

where b is the collision parameter and Ze is the charge on the nucleus. The spectral emissivity of the electron is given by

$$I(\omega) = \frac{e^2}{3\pi\varepsilon c^3}\left[|\dot{v}_{\parallel}(\omega)|^2 + |\dot{v}_{\perp}(\omega)|^2\right]. \tag{1.24}$$

Taking the Fourier transform of Eq. 1.23 we find

$$I(\omega) = \frac{Z^2 e^6 \omega^2}{24\pi^4\varepsilon_0^3 c^3 m_e^2 v^2 \gamma^2 v^2}\left[\frac{1}{\gamma^2}K_0^2\left(\frac{\omega b}{\gamma v}\right) + K_1^2\left(\frac{\omega b}{\gamma v}\right)\right] \tag{1.25}$$

where K_0 and K_1 are modified Bessel functions of order zero and one. In the low frequency limit, $\frac{\omega b}{\gamma v} \ll 1$ the emissivity of the electron can be approximated as

$$\begin{aligned} I(\omega) &= \frac{Z^2 e^6 \omega^2}{24\pi^4\varepsilon_0^3 c^3 m_e^2 v^2 \gamma^2 v^2}\left[\frac{-1}{\gamma^2}\ln^2\left(\frac{\omega b}{\gamma v}\right) + \left(\frac{\gamma v}{\omega b}\right)^2\right], \\ I(\omega) &= \frac{Z^2 e^6}{24\pi^4\varepsilon_0^3 c^3 m_e^2 v^2}\frac{1}{b^2}\left[1 + \frac{1}{\gamma^2}\left(\frac{\omega b}{\gamma b}\right)^2 \ln^2\left(\frac{\omega b}{\gamma v}\right)\right], \\ I(\omega) &\approx \frac{Z^2 e^6}{24\pi^4\varepsilon_0^3 c^3 m_e^2 v^2}\frac{1}{b^2} = K. \end{aligned} \tag{1.26}$$

Integrating over all collision parameters which contribute to the radiation gives

$$I(\omega) = \int_{b_{\min}}^{b_{\min}} 2\pi b N c K db = \frac{Z^2 e^6 N}{12\pi^3\varepsilon_0^3 c^3 m_e^2 v}\ln\Lambda \tag{1.27}$$

where $\Lambda = \frac{b_{\max}}{b_{\min}}$.

In order to get the spectrum for bremsstrahlung radiation at temperature T, we must integrate Eq. 1.27 over a Maxwellian distribution of electrons. The spectral emissivity of a plasma is then given by

$$I(\omega) = \frac{Z^2 e^6 N N_{\mathrm{e}}}{12\sqrt{3}\pi^3 \varepsilon_0^3 c^3 m_{\mathrm{e}}^2} \left(\frac{m_{\mathrm{e}}}{kT}\right)^{\frac{1}{2}} g(\omega, T), \tag{1.28}$$

where N_{e} is the electron density, and $g(\omega, T)$ is known as the Gaunt factor. Alternatively the emissivity can be written in terms of frequency, ν so that

$$k_\nu = \frac{1}{3\pi^2} \left(\frac{\pi}{6}\right)^{\frac{1}{2}} \frac{Z^2 e^6}{\varepsilon_0^3 c^3 m_{\mathrm{e}}^2} \left(\frac{m_{\mathrm{e}}}{kT}\right)^{\frac{1}{2}} g(\nu, T) N N_{\mathrm{e}} \exp\left(-\frac{h\nu}{kT}\right). \tag{1.29}$$

In the case of X-ray emission, the Gaunt factor is

$$g(\nu, T) = \frac{\sqrt{3}}{\pi} \ln\left(\frac{kT}{h\nu}\right). \tag{1.30}$$

The Gaunt factor is a correction factor which takes into account quantum mechanical effects. At high frequencies thermal emission due to bremsstrahlung falls off rapidly. The total energy loss of the plasma is found by integrating Eq. 1.29 over frequency so that

$$-\frac{dE}{dt} = 1.435 \times 10^{-40} Z^2 T^{\frac{1}{2}} \bar{g} N N_e, \tag{1.31}$$

where $\bar{g}$ is the frequency averaged Gaunt factor.

1.3.3 Faraday Rotation

As shown in Sect. 1.3.1 the degree of polarisation of synchrotron radiation from plasma with a uniform magnetic field depends only on the index of the electron population and should be independent of wavelength. Observations of synchrotron sources, such as the Crab nebula, showed a strong dependence on wavelength, with the degree of polarisation sometimes decreasing dramatically with wavelength. In Burn (1966), it was shown that that this depolarisation is often due to Faraday Rotation in the external medium surrounding the source. Faraday Rotation is an important tool for measuring extragalactic magnetic fields (see Sect. 1.1.3). Following Longair (2011) and Rybicki and Lightman (1986) we will derive the equation for Faraday depth.

Maxwell's equations are given by

$$\nabla \cdot \mathbf{E} = \frac{\rho}{\varepsilon} \qquad \nabla \times \mathbf{E} = -\frac{\partial \mathbf{B}}{\partial t}$$
$$\nabla \cdot \mathbf{B} = 0 \qquad \nabla \times \mathbf{B} = \mu \left(\mathbf{J} + \varepsilon \frac{\partial \mathbf{E}}{\partial t} \right) \tag{1.32}$$

where E is the electric field, B is the magnetic field, ρ is the total electric charge density, J is the total electric current density, ε is the permittivity and μ is the permeability. Using the identity

$$\nabla \times (\nabla \times \mathbf{A}) = \nabla (\nabla \cdot \mathbf{A}) - \nabla^2 (\mathbf{A}) \tag{1.33}$$

and Maxwell's equations, we find that

$$\nabla^2 \mathbf{E} = \frac{1}{\mu\varepsilon} \frac{\partial^2 \mathbf{E}}{\partial t^2},$$
$$\nabla^2 \mathbf{B} = \frac{1}{\mu\varepsilon} \frac{\partial^2 \mathbf{B}}{\partial t^2}. \tag{1.34}$$

Equation 1.34 is called the wave equation. One solution to the wave equation is $\mathbf{E} = E_0 e^{i(\mathbf{k}\cdot\mathbf{r}-\omega t)}$ where

$$\varepsilon = \frac{\omega^2}{k^2},$$
$$\mathbf{r} = a_{i_1} \hat{j}_1 + a_{i_2} \hat{j}_2 + a_{i_3} \hat{j}_3,$$
$$\mathbf{k} = k_{i_1} \hat{j}_1 + k_{i_2} \hat{j}_2 + k_{i_3} \hat{j}_3. \tag{1.35}$$

Similarly $\mathbf{B} = B_0 e^{i(\mathbf{k}\cdot\mathbf{r}-\omega t)}$ is a solution. If B_0 and E_0 are constant real vectors then the wave is said to be linearly polarised.

Another solution to the wave equation is $\mathbf{E} = \hat{j}_1 E_0 e^{i\left(ka_{i_3}-\omega t\right)} + \hat{j}_2 E_0 e^{i\left(ka_{i_3}-\omega t \pm \frac{\pi}{2}\right)}$ or $\mathbf{E} = E_0 \left(\hat{j}_1 \pm i \hat{j}_2 \right) e^{i\left(ka_{i_3}-\omega t\right)}$. This wave is described as circularly polarised as the electric and magnetic unit vectors rotate about the direction of propagation. For a phase of $+\frac{\pi}{2}$ the wave is said to be right circularly polarised and for a phase of $-\frac{\pi}{2}$ the wave is said to be left circularly polarised. It is clear that the combination of a right and left circularly polarised wave results in a linearly polarised wave.

We can rewrite Maxwell equations for both linear and circularly polarised waves as

$$i\mathbf{k} \cdot \mathbf{E} = \frac{\rho}{\varepsilon} \qquad i\mathbf{k} \times \mathbf{E} = i\omega \mathbf{B}$$
$$i\mathbf{k} \cdot \mathbf{B} = 0 \qquad i\mathbf{k} \times \mathbf{B} = \mu (\mathbf{J} - i\varepsilon\omega \mathbf{E}) \tag{1.36}$$

Consider a circularly polarised wave propagating through a plasma with magnetic field $\mathbf{B}_0$. Assume the direction of propagation is along the direction of the magnetic field such that $\mathbf{B}_0 = B_0 \hat{j}_3$. The equation of motion for an electron in this plasma is

$$m \frac{d\mathbf{v}}{dt} = -e\mathbf{E} - \frac{e}{c} \mathbf{v} \times \mathbf{B}_0. \tag{1.37}$$

Using the ansatz $\mathbf{v} = v_0 e^{-i\omega t}\left(\hat{j}_1 \pm i\hat{j}_2\right)$ and the fact that $\left(\hat{j}_1 \pm i\hat{j}_2\right) \times \hat{j}_3 = \pm i\left(\hat{j}_1 \pm i\hat{j}_2\right)$ we find that

$$\mathbf{v} = \frac{-ie}{m\left(\omega \pm \omega_{\mathrm{B}}\right)}\mathbf{E}(t) \tag{1.38}$$

where $\omega_{\mathrm{B}} = \frac{eB_0}{mc}$ is the cyclotron frequency which is the frequency of gyration for an electron about the field lines.

From Eq. 1.38 we get the conductivity, $\sigma = \frac{ine^2}{m(\omega\pm\omega_{\mathrm{B}})}$ and the emissivity, $\varepsilon \equiv 1 - \frac{4\pi\sigma}{i\omega} = 1 - \frac{\omega_{\mathrm{p}}^2}{\omega(\omega\pm\omega_{\mathrm{B}})}$. The wave number is then

$$k = \frac{\omega}{c}\sqrt{1 - \frac{\omega_{\mathrm{p}^2}}{\omega\left(\omega \pm \omega_{\mathrm{b}}\right)}}. \tag{1.39}$$

Assuming that $\omega \gg \omega_{\mathrm{B}}$, $\omega \gg \omega_{\mathrm{p}}$ and using the Maclaurin series for $\sqrt{1-x}$ and $\frac{1}{1\pm x}$

$$k = \frac{\omega}{c}\left(1 - \frac{\omega_{\mathrm{p}}^2}{2\omega^2}\left(1 \pm \frac{\omega_{\mathrm{B}}}{\omega}\right)\right). \tag{1.40}$$

The phase of the circularly polarised wave will rotate through an angle $\theta = \int_0^d k ds$ where d is the distance travelled through the plasma. For a linearly polarised wave, the component right and left circularly polarised waves will undergo different phase changes. This leads to a rotation in the angle of polarisation for the linearly polarised wave. The magnitude of this rotation, θ, is the difference in phase changes for the left and right circularly polarised light

$$\begin{aligned}
\theta &= \int_0^d \left(k_{\mathrm{r}} - k_{\mathrm{l}}\right) \mathrm{ds} \\
\theta &= \frac{2\pi e^3}{m^2 c^2 \omega^2}\int_0^d nB_0 \mathrm{ds} \\
\theta &= \frac{e^3\lambda^2}{2\pi m^2 c^4}\int_0^d nB_0 \mathrm{ds} \\
\phi &= \frac{e^3}{2\pi m^2 c^4}\int_0^d nB_0 \mathrm{ds}.
\end{aligned} \tag{1.41}$$

$\phi = \frac{\theta}{\lambda^2}$ is called the Faraday depth. Here we have derived the Faraday depth assuming that the magnetic field is parallel to the direction of propagation. However the equation is still valid if B_0 is replaced with $B_{\|}$.

1.4 Structure of Thesis

In the following chapters I will present low frequency radio observations of several galaxy clusters as well as the giant radio galaxy NGC 6251 which is located in a poor galaxy group. In Chap. 2 I will discuss some of the techniques used in this thesis. In Chap. 3 I present GMRT observations of the massive merging galaxy cluster MACS J22443.3-0935 which was selected for observation based on its highly negative relaxation parameter. I report the discovery of a radio halo in MACSJ2243.3-0935, as well as a new radio relic candidate and discuss the properties these sources. In Chap. 4 I present GMRT observations of 3 dynamically disturbed clusters, also selected for their negative relaxation parameters and place upper limits on the radio halo power in these clusters. In Chap. 5 I present LOFAR observations of the giant radio galaxy NGC 6251. These observations produce the deepest maps of the extended diffuse structure in this source to date. I discuss the contribution these maps make to our understanding of the particle dynamics in the source. I also present the first observations of polarisation in NGC 6251 at 150 MHz and demonstrate the power of RM synthesis in analysing LOFAR polarimetry data.

In this thesis, a ΛCDM cosmology is assumed with $H_0 = 70\,\mathrm{km\,s^{-1}\,Mpc^{-1}}$, $\Omega_\mathrm{m} = 0.3$, $\Omega_\Lambda = 0.7$.

References

Angel JRP, Stockman HS (1980) Optical and infrared polarization of active extragalactic objects. ARA&A 18:321–361. https://doi.org/10.1146/annurev.aa.18.090180.001541

Antonucci R (1993) Unified models for active galactic nuclei and quasars. ARA&A 31:473–521. https://doi.org/10.1146/annurev.aa.31.090193.002353

Baade W (1956) Polarization in the jet of Messier 87. ApJ 123:550–551. https://doi.org/10.1086/146194

Baade W, Minkowski R (1954) Identification of the radio sources in Cassiopeia, Cygnus A, and Puppis A. ApJ 119:206. https://doi.org/10.1086/145812

Best PN, Ker LM, Simpson C, Rigby EE, Sabater J (2014) The cosmic evolution of radio-AGN feedback to z = 1. MNRAS 445:955–969 arXiv:1409.0263

Bîrzan L, McNamara BR, Nulsen PEJ, Carilli CL, Wise MW (2008) Radiative efficiency and content of extragalactic radio sources: toward a universal scaling relation between jet power and radio power. ApJ 686:859–880. https://doi.org/10.1086/591416. arXiv:0806.1929

Blasi P, Colafrancesco S (1999) Cosmic rays, radio halos and nonthermal X-ray emission in clusters of galaxies. Astropart Phys 12(3):169–183. https://doi.org/10.1016/S0927-6505(99)00079-1. http://www.sciencedirect.com/science/article/pii/S0927650599000791

Böhringer H, Pratt GW, Arnaud M, Borgani S, Croston JH, Ponman TJ, Ameglio S, Temple RF, Dolag K (2010) Substructure of the galaxy clusters in the REXCESS sample: observed statistics and comparison to numerical simulations. A&A 514:A32. https://doi.org/10.1051/0004-6361/200913911. arXiv:0912.4667

Bonafede A, Vazza F, Brüggen M, Murgia M, Govoni F, Feretti L, Giovannini G, Ogrean G (2013) Measurements and simulation of Faraday rotation across the Coma radio relic. MNRAS 433:3208–3226. https://doi.org/10.1093/mnras/stt960. arXiv:1305.7228

Bonafede A, Intema HT, Brüggen M, Girardi M, Nonino M, Kantharia N, van Weeren RJ, Röttgering HJA (2014) Evidence for particle re-acceleration in the radio relic in the galaxy cluster PLCKG287.0+32.9. ApJ 785:1. https://doi.org/10.1088/0004-637X/785/1/1. arXiv:1402.1492
Bower RG, Benson AJ, Malbon R, Helly JC, Frenk CS, Baugh CM, Cole S, Lacey CG (2006) Breaking the hierarchy of galaxy formation. MNRAS 370:645–655 arXiv:astro-ph/0511338
Brüggen M, Bykov A, Ryu D, Röttgering H (2012) Magnetic fields, relativistic particles, and shock waves in cluster outskirts. Space Sci Rev 166:187–213. https://doi.org/10.1007/s11214-011-9785-9. arXiv:1107.5223
Brunetti G (2003) Modelling the non-thermal emission from galaxy clusters. In: Bowyer S, Hwang CY (eds) Matter and energy in clusters of galaxies, astronomical society of the pacific conference series, vol 301, p 349. arXiv:astro-ph/0208074
Brunetti G (2004) Particle acceleration and non-thermal emission from galaxy clusters. J Korean Astron Soc 37:493–500. https://doi.org/10.5303/JKAS.2004.37.5.493. arXiv:astro-ph/0412529
Brunetti G (2009) Constraining relativistic protons and magnetic fields in galaxy clusters through radio and γ-ray observations: the case of A2256. A&A 508:599–602. https://doi.org/10.1051/0004-6361/200913177. arXiv:0909.3449
Brunetti G, Jones TW (2014) Cosmic rays in galaxy clusters and their nonthermal emission. Int J Mod Phys D 23(04):1430007. https://doi.org/10.1142/S0218271814300079. arXiv:1401.7519
Brunetti G, Setti G, Feretti L, Giovannini G (2001) Particle reacceleration in the Coma cluster: radio properties and hard X-ray emission. MNRAS 320:365–378. https://doi.org/10.1046/j.1365-8711.2001.03978.x. arXiv:astro-ph/0008518
Brunetti G, Cassano R, Dolag K, Setti G (2009) On the evolution of giant radio halos and their connection with cluster mergers. A&A 507:661–669. https://doi.org/10.1051/0004-6361/200912751. arXiv:0909.2343
Brunetti G, Blasi P, Reimer O, Rudnick L, Bonafede A, Brown S (2012) Probing the origin of giant radio haloes through radio and γ-ray data: the case of the Coma cluster. MNRAS 426:956–968. https://doi.org/10.1111/j.1365-2966.2012.21785.x. arXiv:1207.3025
Buote DA, Tsai JC (1995) Quantifying the morphologies and dynamical evolution of galaxy clusters-I: the method. ApJ 452:522. https://doi.org/10.1086/176326. arXiv:astro-ph/9502002
Burbidge GR (1956) On synchrotron radiation from Messier 87. ApJ 124:416. https://doi.org/10.1086/146237
Burn BJ (1966) On the depolarization of discrete radio sources by Faraday dispersion. MNRAS 133:67. https://doi.org/10.1093/mnras/133.1.67
Carilli CL, Barthel PD (1996) Cygnus A. A&A Rev 7:1–54. https://doi.org/10.1007/s001590050001
Cassano R, Brunetti G (2005) Cluster mergers and non-thermal phenomena: a statistical magneto-turbulent model. MNRAS 357:1313–1329. https://doi.org/10.1111/j.1365-2966.2005.08747.x. arXiv:astro-ph/0412475
Cassano R, Ettori S, Giacintucci S, Brunetti G, Markevitch M, Venturi T, Gitti M (2010) On the connection between giant radio halos and cluster mergers. ApJ 721:L82–L85. https://doi.org/10.1088/2041-8205/721/2/L82. arXiv:1008.3624
Cassano R, Ettori S, Brunetti G, Giacintucci S, Pratt GW, Venturi T, Kale R, Dolag K, Markevitch M (2013) Revisiting scaling relations for giant radio halos in galaxy clusters. ApJ 777:141. https://doi.org/10.1088/0004-637X/777/2/141. arXiv:1306.4379
Cavagnolo KW, Donahue M, Voit GM, Sun M (2009) Intracluster medium entropy profiles for a Chandra archival sample of galaxy clusters. ApJS 182:12–32. https://doi.org/10.1088/0067-0049/182/1/12. arXiv:0902.1802
Cavagnolo KW, McNamara BR, Nulsen PEJ, Carilli CL, Jones C, Bîrzan L (2010) A relationship between AGN jet power and radio power. ApJ 720:1066–1072 arXiv:1006.5699
Chen R, Peng B, Strom RG, Wei J (2011) Group galaxies around giant radio galaxy NGC 6251. MNRAS 412:2433–2444. https://doi.org/10.1111/j.1365-2966.2010.18064.x
Chen R, Peng B, Strom RG, Wei J (2012) Galaxy group around giant radio galaxy NGC 315. MNRAS 420:2715–2726. https://doi.org/10.1111/j.1365-2966.2011.20245.x

Clarke TE, Kronberg PP, Böhringer H (2001) A new radio-x-ray probe of galaxy cluster magnetic fields. ApJ 547:L111–L114. https://doi.org/10.1086/318896. arXiv:astro-ph/0011281
Colafrancesco S (1999) Cosmic Rays and Non-Thermal Emission in Galaxy Clusters. In: Boehringer H, Feretti L, Schuecker P (eds) Diffuse thermal and relativistic plasma in galaxy clusters, p 269. arXiv:astro-ph/9907329
Croston JH, Hardcastle MJ (2014) The particle content of low-power radio galaxies in groups and clusters. MNRAS 438:3310–3321. https://doi.org/10.1093/mnras/stt2436. arXiv:1312.5183
Croston JH, Birkinshaw M, Hardcastle MJ, Worrall DM (2004) X-ray emission from the nuclei, lobes and hot-gas environments of two FR II radio galaxies. MNRAS 353:879–889. https://doi.org/10.1111/j.1365-2966.2004.08118.x. arXiv:astro-ph/0406347
Croston JH, Hardcastle MJ, Harris DE, Belsole E, Birkinshaw M, Worrall DM (2005) An x-ray study of magnetic field strengths and particle content in the lobes of FR II radio sources. ApJ 626:733–747. https://doi.org/10.1086/430170. arXiv:astro-ph/0503203
Croston JH, Hardcastle MJ, Birkinshaw M, Worrall DM, Laing RA (2008) An XMM-Newton study of the environments, particle content and impact of low-power radio galaxies. MNRAS 386:1709–1728. https://doi.org/10.1111/j.1365-2966.2008.13162.x. arXiv:0802.4297
Croton DJ (2006) Evolution in the black hole mass-bulge mass relation: a theoretical perspective. MNRAS 369:1808–1812 arXiv:astro-ph/0512375
Davé R, Cen R, Ostriker JP, Bryan GL, Hernquist L, Katz N, Weinberg DH, Norman ML, O'Shea B (2001) Baryons in the warm-hot intergalactic medium. Astrophys J 552(2):473. http://stacks.iop.org/0004-637X/552/i=2/a=473
Davidson K, Netzer H (1979) The emission lines of quasars and similar objects. Rev Mod Phys 51:715–766. https://doi.org/10.1103/RevModPhys.51.715
De Young DS (2006) The particle content of extragalactic jets. ApJ 648:200–208. https://doi.org/10.1086/505861. arXiv:astro-ph/0605734
Donnert J, Dolag K, Lesch H, Müller E (2009) Cluster magnetic fields from galactic outflows. MNRAS 392:1008–1021. https://doi.org/10.1111/j.1365-2966.2008.14132.x. arXiv:0808.0919
Donnert J, Dolag K, Cassano R, Brunetti G (2010) Radio haloes from simulations and hadronic models-II. The scaling relations of radio haloes. MNRAS 407:1565–1580. https://doi.org/10.1111/j.1365-2966.2010.17065.x. arXiv:1003.0336
Donnert J, Dolag K, Brunetti G, Cassano R (2013) Rise and fall of radio haloes in simulated merging galaxy clusters. Mon Not R Astron Soc 429(4):3564–3569. 10.1093/mnras/sts628. http://mnras.oxfordjournals.org/content/429/4/3564.abstract. http://mnras.oxfordjournals.org/content/429/4/3564.full.pdf+html
Elitzur M (2012) On the unification of active galactic nuclei. ApJ 747:L33. https://doi.org/10.1088/2041-8205/747/2/L33. arXiv:1202.1776
Enßlin TA, Röttgering H (2002) The radio luminosity function of cluster radio halos. A&A 396:83–89. https://doi.org/10.1051/0004-6361:20021382. arXiv:astro-ph/0209218
Enßlin T, Pfrommer C, Miniati F, Subramanian K (2011) Cosmic ray transport in galaxy clusters: implications for radio halos, gamma-ray signatures, and cool core heating. A&A 527:A99. https://doi.org/10.1051/0004-6361/201015652. arXiv:1008.4717
Fabian AC (1994) Cooling flows in clusters of galaxies. ARA&A 32:277–318. https://doi.org/10.1146/annurev.aa.32.090194.001425
Fabian AC (2012) Observational evidence of active galactic nuclei feedback. ARA&A 50:455–489 arXiv:1204.4114
Fanaroff BL, Riley JM (1974) The morphology of extragalactic radio sources of high and low luminosity. MNRAS 167:31P–36P. https://doi.org/10.1093/mnras/167.1.31P
Felten JE, Gould RJ, Stein WA, Woolf NJ (1966) X-rays from the Coma cluster of galaxies. ApJ 146:955–958. https://doi.org/10.1086/148972
Feretti L (2002) Observational properties of diffuse halos in clusters. In: Pramesh Rao A, Swarup G, Gopal-Krishna (eds) The universe at low radio frequencies, IAU symposium, vol 199, p 133

Feretti L (2003) Clusters of galaxies in radio. In: Bowyer S, Hwang CY (eds) Matter and energy in clusters of galaxies, astronomical society of the pacific conference series, vol 301, p 143. arXiv:astro-ph/0301576

Feretti L, Dallacasa D, Govoni F, Giovannini G, Taylor GB, Klein U (1999) The radio galaxies and the magnetic field in Abell 119. A&A 344:472–482 arXiv:astro-ph/9902019

Feretti L, Giovannini G, Govoni F, Murgia M (2012) Clusters of galaxies: observational properties of the diffuse radio emission. Astron Astrophys Rev 20(1):54. https://doi.org/10.1007/s00159-012-0054-z. arXiv:1205.1919v1

Fujita Y, Takizawa M, Sarazin CL (2003) Nonthermal emissions from particles accelerated by turbulence in clusters of galaxies. ApJ 584:190–202. https://doi.org/10.1086/345599. arXiv:astro-ph/0210320

Gaspari M, Brighenti F, Ruszkowski M (2013) Solving the cooling flow problem through mechanical AGN feedback. Astron Nachr 334:394. https://doi.org/10.1002/asna.201211865. arXiv:1209.3305

Gizani NAB, Leahy JP (2003) A multiband study of Hercules A-II. Multifrequency very large array imaging. MNRAS 342:399–421. https://doi.org/10.1046/j.1365-8711.2003.06469.x. arXiv:astro-ph/0305600

Godfrey LEH, Shabala SS (2016) Mutual distance dependence drives the observed jet-power-radio-luminosity scaling relations in radio galaxies. MNRAS 456:1172–1184. https://doi.org/10.1093/mnras/stv2712. arXiv:1511.06007

Govoni F, Feretti L (2004) Magnetic fields in clusters of galaxies. Int J Mod Phys D 13:1549–1594. https://doi.org/10.1142/S0218271804005080. arXiv:astro-ph/0410182

Govoni F, Enßlin TA, Feretti L, Giovannini G (2001) A comparison of radio and X-ray morphologies of four clusters of galaxies containing radio halos. A&A 369:441–449. https://doi.org/10.1051/0004-6361:20010115. arXiv:astro-ph/0101418

Govoni F, Murgia M, Feretti L, Giovannini G, Dolag K, Taylor GB (2006) The intracluster magnetic field power spectrum in Abell 2255. A&A 460:425–438. https://doi.org/10.1051/0004-6361:20065964. arXiv:astro-ph/0608433

Govoni F, Dolag K, Murgia M, Feretti L, Schindler S, Giovannini G, Boschin W, Vacca V, Bonafede A (2010) Rotation measures of radio sources in hot galaxy clusters. A&A 522:A105. https://doi.org/10.1051/0004-6361/200913665. arXiv:1007.5207

Graham I (1970) Observations of M87 at 5 GHz. MNRAS 149:319–339. https://doi.org/10.1093/mnras/149.4.319

Hardcastle MJ, Worrall DM (2000) The environments of FRII radio sources. MNRAS 319:562–572. https://doi.org/10.1046/j.1365-8711.2000.03883.x. arXiv:astro-ph/0007260

Hardcastle MJ, Worrall DM, Birkinshaw M, Laing RA, Bridle AH (2002) A Chandra observation of the X-ray environment and jet of 3C 31. MNRAS 334:182–192. https://doi.org/10.1046/j.1365-8711.2002.05513.x. arXiv:astro-ph/0203374

Hardcastle MJ, Worrall DM, Kraft RP, Forman WR, Jones C, Murray SS (2003) Radio and x-ray observations of the jet in Centaurus A. ApJ 593:169–183. https://doi.org/10.1086/376519. arXiv:astro-ph/0304443

Hardcastle MJ, Croston JH, Kraft RP (2007) A Chandra study of particle acceleration in the multiple hot spots of nearby radio galaxies. ApJ 669:893–904. https://doi.org/10.1086/521696. arXiv:0707.2865

Hargrave PJ, Ryle M (1974) Observations of Cygnus A with the 5-km radio telescope. MNRAS 166:305–327. https://doi.org/10.1093/mnras/166.2.305

Heckman TM, Best PN (2014) The coevolution of galaxies and supermassive black holes: insights from surveys of the contemporary universe. ARA&A 52:589–660. https://doi.org/10.1146/annurev-astro-081913-035722. arXiv:1403.4620

Hook IM, McMahon RG, Boyle BJ, Irwin MJ (1994) The variability of optically selected quasars. MNRAS 268:305. https://doi.org/10.1093/mnras/268.2.305

Hudson DS, Mittal R, Reiprich TH, Nulsen PEJ, Andernach H, Sarazin CL (2010) What is a cool-core cluster? A detailed analysis of the cores of the X-ray flux-limited HIFLUGCS cluster sample. A&A 513:A37. https://doi.org/10.1051/0004-6361/200912377. arXiv:0911.0409

Ineson J, Croston JH, Hardcastle MJ, Mingo B (2017) A representative survey of the dynamics and energetics of FR II radio galaxies. MNRAS 467:1586–1607. https://doi.org/10.1093/mnras/stx189. arXiv:1701.05612

Ishwara-Chandra CH, Saikia DJ (1999) Giant radio sources. MNRAS 309:100–112. https://doi.org/10.1046/j.1365-8711.1999.02835.x. arXiv:astro-ph/9902252

Isobe N, Koyama S (2015) X-ray measurement of electron and magnetic-field energy densities in the west lobe of the giant radio galaxy 3C 236. PASJ 67:77. https://doi.org/10.1093/pasj/psv046. arXiv:1505.02769

Jeltema TE, Profumo S (2011) Implications of fermi observations for hadronic models of radio halos in clusters of galaxies. ApJ 728:53. https://doi.org/10.1088/0004-637X/728/1/53. arXiv:1006.1648

Jennison RC, Das Gupta MK (1953) Fine structure of the extra-terrestrial radio source Cygnus I. Nature 172:996–997. https://doi.org/10.1038/172996a0

Jones FC, Ellison DC (1991) The plasma physics of shock acceleration. Space Sci. Rev. 58:259–346. https://doi.org/10.1007/BF01206003

Kang H, Ryu D, Jones TW (2012) Diffusive shock acceleration simulations of radio relics. ApJ 756:97. https://doi.org/10.1088/0004-637X/756/1/97. arXiv:1205.1895

Kawakatu N, Kino M, Takahara F (2016) Evidence for a significant mixture of electron/positron pairs in FRII jets constrained by cocoon dynamics. MNRAS 457:1124–1136. https://doi.org/10.1093/mnras/stw010. arXiv:1601.00771

Khachikian EY, Weedman DW (1974) An atlas of Seyfert galaxies. ApJ 192:581–589. https://doi.org/10.1086/153093

King I (1962) The structure of star clusters-I: an empirical density law. AJ 67:471. https://doi.org/10.1086/108756

Kokotanekov G, Wise M, Heald GH, McKean JP, Bîrzan L, Rafferty DA, Godfrey LEH, de Vries M, Intema HT, Broderick JW, Hardcastle MJ, Bonafede A, Clarke AO, van Weeren RJ, Röttgering HJA, Pizzo R, Iacobelli M, Orrú E, Shulevski A, Riseley CJ, Breton RP, Nikiel-Wroczyński B, Sridhar SS, Stewart AJ, Rowlinson A, van der Horst AJ, Harwood JJ, Gürkan G, Carbone D, Pandey-Pommier M, Tasse C, Scaife AMM, Pratley L, Ferrari C, Croston JH, Pandey VN, Jurusik W, Mulcahy DD (2017) LOFAR MSSS: the scaling relation between AGN cavity power and radio luminosity at low radio frequencies. arXiv:1706.00225

Komberg BV, Pashchenko IN (2009) Giant radio galaxies: old long-lived quasars? Astron Rep 53:1086–1100. https://doi.org/10.1134/S1063772909120026. arXiv:0901.3721

Kravtsov AV, Borgani S (2012) Formation of galaxy clusters. ARA&A 50:353–409. https://doi.org/10.1146/annurev-astro-081811-125502. arXiv:1205.5556

Laing RA, Bridle AH (2002) Dynamical models for jet deceleration in the radio galaxy 3C 31. MNRAS 336:1161–1180. https://doi.org/10.1046/j.1365-8711.2002.05873.x. arXiv:astro-ph/0207427

Laing RA, Canvin JR, Cotton WD, Bridle AH (2006) Multifrequency observations of the jets in the radio galaxy NGC315. MNRAS 368:48–64. https://doi.org/10.1111/j.1365-2966.2006.10099.x. arXiv:astro-ph/0601660

Laing RA, Bridle AH, Parma P, Feretti L, Giovannini G, Murgia M, Perley RA (2008) Multifrequency VLA observations of the FR I radio galaxy 3C 31: morphology, spectrum and magnetic field. MNRAS 386:657–672. https://doi.org/10.1111/j.1365-2966.2008.13091.x. arXiv:0803.2597

Lara L, Cotton WD, Feretti L, Giovannini G, Marcaide JM, Márquez I, Venturi T (2001) A new sample of large angular size radio galaxies. I. The radio data. A&A 370:409–425. https://doi.org/10.1051/0004-6361:20010254. arXiv:astro-ph/0102034

Li H, Lapenta G, Finn JM, Li S, Colgate SA (2006) Modeling the large-scale structures of astrophysical jets in the magnetically dominated limit. ApJ 643:92–100. https://doi.org/10.1086/501499. arXiv:astro-ph/0604469

Longair MS (2011) High energy astrophysics

Machalski J, Jamrozy M (2006) The new sample of giant radio sources. III. Statistical trends and correlations. A&A 454:95–102. https://doi.org/10.1051/0004-6361:20054673. arXiv:astro-ph/0605011

Machalski J, Jamrozy M, Stawarz Ł, Kozieł-Wierzbowska D (2011) Understanding giant radio galaxy J1420–0545: large-scale morphology, environment, and energetics. ApJ 740:58. https://doi.org/10.1088/0004-637X/740/2/58. arXiv:1107.5449

Machalski J, Chyzy KT, Jamrozy M (2002) On the time evolution of giant radio galaxies. ArXiv Astrophys e-prints. arXiv:astro-ph/0210546

Mack KH, Klein U, O'Dea CP, Willis AG (1997) Multi-frequency radio continuum mapping of giant radio galaxies. A&A S 123. https://doi.org/10.1051/aas:1997166

Mack KH, Klein U, O'Dea CP, Willis AG, Saripalli L (1998) Spectral indices, particle ages, and the ambient medium of giant radio galaxies. A&A 329:431–442

MacLeod CL, Ivezić Ž, Kochanek CS, Kozłowski S, Kelly B, Bullock E, Kimball A, Sesar B, Westman D, Brooks K, Gibson R, Becker AC, de Vries WH (2010) Modeling the time variability of SDSS stripe 82 quasars as a damped random walk. ApJ 721:1014–1033. https://doi.org/10.1088/0004-637X/721/2/1014. arXiv:1004.0276

Maiolino R, Rieke GH (1995) Low-luminosity and obscured Seyfert nuclei in nearby galaxies. ApJ 454:95. https://doi.org/10.1086/176468

Malarecki JM, Jones DH, Saripalli L, Staveley-Smith L, Subrahmanyan R (2015) Giant radio galaxies-II. Tracers of large-scale structure. MNRAS 449:955–986. https://doi.org/10.1093/mnras/stv273. arXiv:1502.03954

Mann AW, Ebeling H (2012) X-ray-optical classification of cluster mergers and the evolution of the cluster merger fraction. Mon Not R Astron Soc 420(3):2120–2138. 10.1111/j.1365-2966.2011.20170.x. http://mnras.oxfordjournals.org/content/420/3/2120.abstract. http://mnras.oxfordjournals.org/content/420/3/2120.full.pdf+html

Marchegiani P, Perola GC, Colafrancesco S (2007) Testing the cosmic ray content in galaxy clusters. A&A 465:41–49. https://doi.org/10.1051/0004-6361:20065977. arXiv:astro-ph/0701592

Markevitch M (2012) Intergalactic shock fronts. World Scientific, pp 397–410. https://doi.org/10.1142/9789814374552_0018

Markevitch M, Govoni F, Brunetti G, Jerius D (2005) Bow shock and radio halo in the merging cluster A520. ApJ 627:733–738. https://doi.org/10.1086/430695. arXiv:astro-ph/0412451

Matthews TA, Sandage AR (1963) Optical identification of 3C 48, 3C 196, and 3C 286 with stellar objects. ApJ 138:30. https://doi.org/10.1086/147615

McNamara BR, Nulsen PEJ (2007) Heating hot atmospheres with active galactic nuclei. ARA&A 45:117–175. https://doi.org/10.1146/annurev.astro.45.051806.110625. arXiv:0709.2152

Miniati F, Jones TW, Kang H, Ryu D (2001) Cosmic-ray electrons in groups and clusters of galaxies: primary and secondary populations from a numerical cosmological simulation. Astrophys J 562(1):233. http://stacks.iop.org/0004-637X/562/i=1/a=233

Mohr JJ, Evrard AE, Fabricant DG, Geller MJ (1995) Cosmological constraints from observed cluster x-ray morphologies. ApJ 447:8. https://doi.org/10.1086/175852. arXiv:astro-ph/9501011

Morganti R, Fanti R, Gioia IM, Harris DE, Parma P, de Ruiter H (1988) Low luminosity radio galaxies-effects of gaseous environment. A&A 189:11–26

Mulcahy DD, Horneffer A, Beck R, Heald G, Fletcher A, Scaife A, Adebahr B, Anderson JM, Bonafede A, Brüggen M, Brunetti G, Chyy KT, Conway J, Dettmar RJ, Enßlin T, Haverkorn M, Horellou C, Iacobelli M, Israel FP, Junklewitz H, Jurusik W, Köhler J, Kuniyoshi M, Orrú E, Paladino R, Pizzo R, Reich W, Röttgering HJA (2014) The nature of the low-frequency emission of M 51. First observations of a nearby galaxy with LOFAR. A&A 568:A74, https://doi.org/10.1051/0004-6361/201424187. arXiv:1407.1312

Mulchaey JS (2000) X-ray properties of groups of galaxies. ARA&A 38:289–335. https://doi.org/10.1146/annurev.astro.38.1.289. arXiv:astro-ph/0009379
Nakamura M, Li H, Li S (2006) Structure of magnetic tower jets in stratified atmospheres. ApJ 652:1059–1067. https://doi.org/10.1086/508338. arXiv:astro-ph/0608326
Netzer H (2015) Revisiting the unified model of active galactic nuclei. ARA&A 53:365–408. https://doi.org/10.1146/annurev-astro-082214-122302. arXiv:1505.00811
Nicastro F, Mathur S, Elvis M (2008) Missing baryons and the warm-hot intergalactic medium. Science 319:55. https://doi.org/10.1126/science.1151400. arXiv:0712.2375
Orrù E, van Velzen S, Pizzo RF, Yatawatta S, Paladino R, Iacobelli M, Murgia M, Falcke H, Morganti R, de Bruyn AG, Ferrari C, Anderson J, Bonafede A, Mulcahy D, Asgekar A, Avruch IM, Beck R, Bell ME, van Bemmel I, Bentum MJ, Bernardi G, Best P, Breitling F, Broderick JW, Brüggen M, Butcher HR, Ciardi B, Conway JE, Corstanje A, de Geus E, Deller A, Duscha S, Eislöffel J, Engels D, Frieswijk W, Garrett MA, Grießmeier J, Gunst AW, Hamaker JP, Heald G, Hoeft M, van der Horst AJ, Intema H, Juette E, Kohler J, Kondratiev VI, Kuniyoshi M, Kuper G, Loose M, Maat P, Mann G, Markoff S, McFadden R, McKay-Bukowski D, Miley G, Moldon J, Molenaar G, Munk H, Nelles A, Paas H, Pandey-Pommier M, Pandey VN, Pietka G, Polatidis AG, Reich W, Röttgering H, Rowlinson A, Scaife A, Schoenmakers A, Schwarz D, Serylak M, Shulevski A, Smirnov O, Steinmetz M, Stewart A, Swinbank J, Tagger M, Tasse C, Thoudam S, Toribio MC, Vermeulen R, Vocks C, van Weeren RJ, Wijers RAMJ, Wise MW, Wucknitz O (2015) Wide-field LOFAR imaging of the field around the double-double radio galaxy B1834+620. A fresh view on a restarted AGN and doubeltjes. A&A 584:A112. https://doi.org/10.1051/0004-6361/201526501. arXiv:1510.00577
Owen FN, Ledlow MJ (1994) The FRI/II break and the bivariate luminosity function in Abell clusters of galaxies. In: Bicknell GV, Dopita MA, Quinn PJ (eds) The physics of active galaxies. Astronomical Society of the Pacific Conference Series, vol 54, p 319
Owen FN (1993) Steps toward a radio H-R diagram. In: Röser HJ, Meisenheimer K (eds) Jets in extragalactic radio sources, vol 421. Lecture Notes in Physics. Springer, Berlin, p 273. https://doi.org/10.1007/3-540-57164-7_104
Padovani P, Giommi P (1995) A sample-oriented catalogue of Bl-lacertae objects. MNRAS 277:1477. https://doi.org/10.1093/mnras/277.4.1477. arXiv:astro-ph/9511065
Paul S, John RS, Gupta P, Kumar H (2017) Understanding 'galaxy groups' as a unique structure in the universe. MNRAS 471:2–11. https://doi.org/10.1093/mnras/stx1488. arXiv:1706.01916
Perley RA, Bridle AH, Willis AG (1984) High-resolution VLA observations of the radio jet in NGC 6251. ApJS 54:291–334. https://doi.org/10.1086/190931
Peterson JR, Fabian AC (2006) X-ray spectroscopy of cooling clusters. Phys Rep 427:1–39. https://doi.org/10.1016/j.physrep.2005.12.007. arXiv:astro-ph/0512549
Peterson BM, Crenshaw DM, Meyers KA, Byard PL, Foltz CB (1984) Variability of the emission-line spectra and optical continua of Seyfert galaxies. II. ApJ 279:529–540. https://doi.org/10.1086/161917
Peterson BM, Wanders I, Bertram R, Hunley JF, Pogge RW, Wagner RM (1998) Optical continuum and emission-line variability of Seyfert 1 galaxies. ApJ 501:82–93. https://doi.org/10.1086/305813. arXiv:astro-ph/9802104
Petrosian V (2001) On the nonthermal emission and acceleration of electrons in coma and other clusters of galaxies. ApJ 557:560–572. https://doi.org/10.1086/321557. arXiv:astro-ph/0101145
Pirya A, Saikia DJ, Singh M, Chandola HC (2012) A study of the environments of large radio galaxies using SDSS. MNRAS 426:758–763. https://doi.org/10.1111/j.1365-2966.2012.21656.x. arXiv:1207.1566
Ponman TJ, Cannon DB, Navarro JF (1999) The thermal imprint of galaxy formation on X-ray clusters. Nature 397:135–137. https://doi.org/10.1038/16410. arXiv:astro-ph/9810359
Ponman TJ, Sanderson AJR, Finoguenov A (2003) The Birmingham-CfA cluster scaling project-III. Entropy and similarity in galaxy systems. MNRAS 343:331–342. https://doi.org/10.1046/j.1365-8711.2003.06677.x. arXiv:astro-ph/0304048
Rybicki GB, Lightman AP (1986) Radiative processes in astrophysics

Ryle M, Elsmore B, Neville AC (1965) High-resolution observations of the radio sources in Cygnus and Cassiopeia. Nature 205:1259–1262. https://doi.org/10.1038/2051259a0
Safouris V, Subrahmanyan R, Bicknell GV, Saripalli L (2009) MRCB0319-454: probing the large-scale structure with a giant radio galaxy. MNRAS 393:2–20. https://doi.org/10.1111/j.1365-2966.2008.14181.x. arXiv:0812.2052
Sanderson AJR, Ponman TJ, O'Sullivan E (2006) A statistically selected Chandra sample of 20 galaxy clusters-I. Temperature and cooling time profiles. MNRAS 372:1496–1508. https://doi.org/10.1111/j.1365-2966.2006.10956.x. arXiv:astro-ph/0608423
Sanderson AJR, Edge AC, Smith GP (2009a) LoCuSS: the connection between brightest cluster galaxy activity, gas cooling and dynamical disturbance of X-ray cluster cores. MNRAS 398:1698–1705. https://doi.org/10.1111/j.1365-2966.2009.15214.x. arXiv:0906.1808
Sanderson AJR, O'Sullivan E, Ponman TJ (2009b) A statistically selected Chandra sample of 20 galaxy clusters-II. Gas properties and cool core/non-cool core bimodality. MNRAS 395:764–776. https://doi.org/10.1111/j.1365-2966.2009.14613.x. arXiv:0902.1747
Sarazin CL (1988) X-ray emission from clusters of galaxies
Saripalli L, Patnaik AR, Porcas RW, Graham DA (1997) Nuclear radio emission in megaparsec-size radio galaxies. A&A 328:78–82
Saripalli L, Hunstead RW, Subrahmanyan R, Boyce E (2005) A complete sample of megaparsec-sized double radio sources from the Sydney University Molonglo Sky Survey. AJ 130:896–922. https://doi.org/10.1086/432507. arXiv:astro-ph/0507055
Schmidt M (1963) 3C 273: a star-like object with large red-shift. Nature 197:1040. https://doi.org/10.1038/1971040a0
Schoenmakers AP, Mack KH, Lara L, Röttgering HJA, de Bruyn AG, van der Laan H, Giovannini G (1998) WNB 0313+683: analysis of a newly discovered giant radio galaxy. A&A 336:455–478 arXiv:astro-ph/9805356
Schoenmakers AP, Mack KH, de Bruyn AG, Röttgering HJA, Klein U, van der Laan H (2000) A new sample of giant radio galaxies from the WENSS survey. II. A multi-frequency radio study of a complete sample: properties of the radio lobes and their environment. A&A S 146:293–322. https://doi.org/10.1051/aas:2000267. arXiv:astro-ph/0008246
Shelton DL, Hardcastle MJ, Croston JH (2011) The dynamics and environmental impact of 3C 452. MNRAS 418:811–819. https://doi.org/10.1111/j.1365-2966.2011.19533.x. arXiv:1108.3753
Sijacki D, Springel V, Di Matteo T, Hernquist L (2007) A unified model for AGN feedback in cosmological simulations of structure formation. MNRAS 380:877–900 arXiv:0705.2238
Smith BD, Hallman EJ, Shull JM, O'Shea BW (2011) The nature of the warm/hot intergalactic medium. I. Numerical methods, convergence, and O VI absorption. ApJ 731:6. https://doi.org/10.1088/0004-637X/731/1/6. arXiv:1009.0261
Sommer MW, Basu K, Intema H, Pacaud F, Bonafede A, Babul A, Bertoldi F (2017) Mpc-scale diffuse radio emission in two massive cool-core clusters of galaxies. MNRAS 466:996–1009. https://doi.org/10.1093/mnras/stw3015. arXiv:1610.07875
Subrahmanyan R, Saripalli L, Safouris V, Hunstead RW (2008) On the relationship between a giant radio galaxy MSH 05–22 and the ambient large-scale galaxy structure. ApJ 677:63–78. https://doi.org/10.1086/529007. arXiv:0801.3910
Sun M (2012) Hot gas in galaxy groups: recent observations. New J Phys 14(4):045004. https://doi.org/10.1088/1367-2630/14/4/045004. arXiv:1203.4228
Tully RB (2015) Galaxy groups: a 2MASS catalog. AJ 149:171. https://doi.org/10.1088/0004-6256/149/5/171. arXiv:1503.03134
Tully RB (1987) Nearby groups of galaxies. II-an all-sky survey within 3000 km per second. ApJ 321:280–304. https://doi.org/10.1086/165629
Turland BD (1975) Observations of M87 at 5 GHz with the 5-km telescope. MNRAS 170:281–294. https://doi.org/10.1093/mnras/170.2.281
van Weeren RJ, Andrade-Santos F, Dawson WA, Golovich N, Lal DV, Kang H, Ryu D, Brìggen M, Ogrean GA, Forman WR, Jones C, Placco VM, Santucci RM, Wittman D, Jee MJ, Kraft

RP, Sobral D, Stroe A, Fogarty K (2017) The case for electron re-acceleration at galaxy cluster shocks. Nat Astron 1:0005. https://doi.org/10.1038/s41550-016-0005. arXiv:1701.01439

van Weeren RJ, Brunetti G, Brüggen M, Andrade-Santos F, Ogrean GA, Williams WL, Röttgering HJA, Dawson WA, Forman WR, de Gasperin F, Hardcastle MJ, Jones C, Miley GK, Rafferty DA, Rudnick L, Sabater J, Sarazin CL, Shimwell TW, Bonafede A, Best PN, Bîrzan L, Cassano R, Chyy KT, Croston JH, Dijkema TJ, Enßlin T, Ferrari C, Heald G, Hoeft M, Horellou C, Jarvis MJ, Kraft RP, Mevius M, Intema HT, Murray SS, Orrú E, Pizzo R, Sridhar SS, Simionescu A, Stroe A, van der Tol S, White GJ (2016) LOFAR, VLA, and Chandra observations of the toothbrush galaxy cluster. ApJ 818:204. https://doi.org/10.3847/0004-637X/818/2/204. arXiv:1601.06029

Vanden Berk DE, Wilhite BC, Kron RG, Anderson SF, Brunner RJ, Hall PB, Ivezić Ž, Richards GT, Schneider DP, York DG, Brinkmann JV, Lamb DQ, Nichol RC, Schlegel DJ (2004) The ensemble photometric variability of ~25,000 quasars in the sloan digital sky survey. ApJ 601:692–714. https://doi.org/10.1086/380563. arXiv:astro-ph/0310336

Vazza F (2016) The quest for extragalactic magnetic fields. In: Proceedings of the neutrino oscillation workshop, p 47. arXiv:1611.00043

Vazza F, Brüggen M (2014) Do radio relics challenge diffusive shock acceleration? MNRAS 437:2291–2296. https://doi.org/10.1093/mnras/stt2042. arXiv:1310.5707

Vazza F, Brüggen M, Wittor D, Gheller C, Eckert D, Stubbe M (2016) Constraining the efficiency of cosmic ray acceleration by cluster shocks. MNRAS 459:70–83. https://doi.org/10.1093/mnras/stw584. arXiv:1603.02688

Venturi T, Rossetti M, Brunetti G, Farnsworth D, Gastaldello F, Giacintucci S, Lal DV, Rudnick L, Shimwell TW, Eckert D, Molendi S, Owers M (2017) The two-component giant radio halo in the galaxy cluster Abell 2142. A&A 603:A125. https://doi.org/10.1051/0004-6361/201630014. arXiv:1703.06802

Voit GM (2005) Tracing cosmic evolution with clusters of galaxies. Rev Mod Phys 77:207–258. https://doi.org/10.1103/RevModPhys.77.207. arXiv:astro-ph/0410173

Wen ZL, Han JL (2013) Substructure and dynamical state of 2092 rich clusters of galaxies derived from photometric data. MNRAS 436:275–293. https://doi.org/10.1093/mnras/stt1581. arXiv:1307.0568

Willson MAG (1970) Radio observations of the cluster of galaxies in Coma Berenices-the 5C4 survey. MNRAS 151:1–44. https://doi.org/10.1093/mnras/151.1.1

Worrall DM, Birkinshaw M (2000) X-ray-emitting atmospheres of B2 radio galaxies. ApJ 530:719–732. https://doi.org/10.1086/308411. arXiv:astro-ph/9910141

Yuan ZS, Han JL, Wen ZL (2015) The scaling relations and the fundamental plane for radio halos and relics of galaxy clusters. ApJ 813(1):77. http://stacks.iop.org/0004-637X/813/i=1/a=77

Zandanel F, Pfrommer C, Prada F (2014) On the physics of radio haloes in galaxy clusters: scaling relations and luminosity functions. MNRAS 438:124–144. https://doi.org/10.1093/mnras/stt2250. arXiv:1311.4795

Zensus JA (1989) Superluminal motion in quasars and Bl-lacertae objects. In: Maraschi L, Maccacaro T, Ulrich MH (eds) BL Lac objects, vol 334. Lecture Notes in Physics. Springer, Berlin, p 3. https://doi.org/10.1007/BFb0031137

Chapter 2
Techniques and Data Calibration

2.1 Radio Interferometry

2.1.1 *Why Is Interferometry Needed*

The resolution of a radio telescope is determined by the equation $\theta = 1.22\frac{\lambda}{D}$, where λ is the observing wavelength and D is the diameter of the dish. Increasing the resolution of single dish radio astronomy requires the construction of larger and larger dishes, which quickly becomes both impractical and expensive. The largest radio dish constructed is the Five-hundred-meter Aperture Spherical radio Telescope (FAST) radio dish in China, with an effective diameter of 300 m and a resolution of 2.9 arcmin at L-Band (Nan et al. 2011).

The technique of radio interferometry was developed in order to overcome this limitation. This technique uses an array of smaller radio telescopes with precisely known locations, as well as accurate signal arrival times, to synthesise a much larger dish with a resolution that is given by $\theta = 1.22\frac{\lambda}{B}$, where B is the longest baseline, or distance between individual antennae, measured in meters.

In the following sections I will describe some of the basic principles in radio interferometry as well as the methods involved in calibrating interferometric data. For a complete and thorough description of radio interforometry see Thompson et al. (2017).

2.1.2 *Basic Interferometry*

Consider an incoherent source located in a distant plane. The radiated field is measured by antenna 1 at $\mathbf{r}_1$ and antenna 2 at $\mathbf{r}_2$. First consider a single element in the source, S, located at $\mathbf{R}$. The source distance is $R = |\mathbf{R}|$ in the direction $\hat{s} = \frac{\mathbf{R}}{R}$. The distance between antenna 1 and antenna 2 is $B = |\mathbf{r}_1 - \mathbf{r}_2|$. The field measured at antenna 1 is given by

T. Cantwell, *Low Frequency Radio Observations of Galaxy Clusters and Groups*, Springer Theses, https://doi.org/10.1007/978-3-319-97976-2_2

$$\mathbf{E}_1(\mathbf{r}_1, t) = \mathcal{E}\left(\hat{s}, t - \frac{R_1}{c}\right)\frac{\mathrm{e}^{-2i\pi\nu\left(t-\frac{R_1}{c}\right)}}{R_1}, \tag{2.1}$$

where $\mathcal{E}\left(\hat{s}, t - \frac{R_1}{c}\right)$ is the complex electric field amplitude vector emitted by element S in the direction of $\mathbf{r}_1$. Similarly the field measured at antenna 2 is given by

$$\mathbf{E}_2(\mathbf{r}_2, t) = \mathcal{E}\left(\hat{s}, t - \frac{R_2}{c}\right)\frac{\mathrm{e}^{-2i\pi\nu\left(t-\frac{R_2}{c}\right)}}{R_2}. \tag{2.2}$$

The exponential terms in Eqs. 2.1 and 2.2 represent the phase change in traversing the paths from element S to antenna 1 and antenna 2.

The complex cross correlation of the field voltages at antenna 1 and antenna 2 is then

$$\begin{aligned}\langle\mathbf{E}_1(\mathbf{r}_1, t)\,\mathbf{E}_2^*(\mathbf{r}_2, t)\rangle &= \langle\mathcal{E}\left(\hat{s}, t - \frac{R_1}{c}\right)\mathcal{E}^*\left(\hat{s}, t - \frac{R_2}{c}\right)\rangle\frac{\mathrm{e}^{-2i\pi\nu\left(t-\frac{R_1}{c}\right)}\mathrm{e}^{2i\pi\nu\left(t-\frac{R_2}{c}\right)}}{R_1R_2}\\ &= \langle\mathcal{E}\left(\hat{s}, t\right)\mathcal{E}^*\left(\hat{s}, t - \frac{R_1 - R_2}{c}\right)\rangle\frac{\mathrm{e}^{-2i\pi\nu\left(\frac{R_1-R_2}{c}\right)}}{R_1R_2}.\end{aligned} \tag{2.3}$$

If the $\frac{R_1-R_2}{c}$ is small compared with the reciprical reciever bandwidth and if the distance to the source is much greater than the distance between antenna 1 and antenna 2 then

$$\langle\mathbf{E}_1(\mathbf{r}_1, t)\,\mathbf{E}_2^*(\mathbf{r}_2, t)\rangle = \langle\mathcal{E}\left(\hat{s}, t\right)\mathcal{E}^*\left(\hat{s}, t\right)\rangle\frac{\mathrm{e}^{-2i\pi\nu\left(\frac{R_1-R_2}{c}\right)}}{R^2}. \tag{2.4}$$

The term $\langle\mathcal{E}\left(\hat{s}, t\right)\mathcal{E}^*\left(\hat{s}, t\right)\rangle$ is the time averaged intensity, $I(\hat{s})$, of the source so that

$$\langle\mathbf{E}_1(\mathbf{r}_1, t)\,\mathbf{E}_2^*(\mathbf{r}_2, t)\rangle = I(\hat{s})\frac{\mathrm{e}^{-2i\pi\nu\frac{R_1-R_2}{c}}}{R^2}. \tag{2.5}$$

The mutual coherence function, $\Gamma(\mathbf{r}_1, \mathbf{r}_2, 0)$, for the whole source is obtained by integrating Eq. 2.5 over the source so that

$$\Gamma(\mathbf{r}_1, \mathbf{r}_2, 0) = \int_{\text{source}}\frac{I(\hat{s})\mathrm{e}^{-2i\pi\nu\hat{s}\cdot\left(\frac{\mathbf{r}_1-\mathbf{r}_2}{c}\right)}}{R^2}ds, \tag{2.6}$$

where ds represents an element of the source. In radio astronomy the convention is to place the phase reference point at the centre of the field of view such that

$$\mathcal{V}(u, v, w) = \mathrm{e}^{2i\pi\hat{s}_0\cdot\mathbf{B}_\lambda}\Gamma(\mathbf{r}_1, \mathbf{r}_2, 0), \tag{2.7}$$

where $\mathcal{V}(u, v, w)$ is the complex visibility, $\hat{s}_0$ is the unit vector in the direction of the phase centre and $\mathbf{B}_\lambda = \frac{\mathbf{B}}{\lambda}$.

If we define our coordinate system such that $\hat{x}$ is to the east, $\hat{y}$ is to the north and $\hat{z}$ is towards the phase centre then

$$\begin{aligned} \hat{s} &= l\hat{x} + m\hat{y} + n\hat{z} \\ \hat{s_0} &= \hat{z} \\ \mathbf{B}_\lambda &= u\hat{x} + v\hat{y} + w\hat{z} \\ d\Omega &= \frac{dldm}{\sqrt{1 - l^2 - m^2}} \end{aligned} \tag{2.8}$$

where l, m, and n are the directional cosines and $n = \sqrt{1 - l^2 - m^2}$. We can rewrite Eq. 2.7 as

$$\begin{aligned} \mathcal{V}(u, v, w) &= \int_{\text{source}} I(\hat{s}) \mathrm{e}^{-2i\pi \hat{s}\cdot\mathbf{B}_\lambda} \mathrm{e}^{2i\pi \hat{s_0}\cdot\mathbf{B}_\lambda} d\Omega \\ &= \int_{\text{source}} I(\hat{s}) \mathrm{e}^{-2i\pi (l\hat{x}+m\hat{y}+(n-1)\hat{z})\cdot(u\hat{x}+v\hat{y}+w\hat{z})} d\Omega \\ &= \int_{\text{source}} \frac{I(\hat{s}) \mathrm{e}^{-2i\pi(ul+mv+(n-1)w)}}{n} dldm. \end{aligned} \tag{2.9}$$

For small enough l and m, $n \approx 1$ so that

$$\mathcal{V}(u, v) = \int_{\text{source}} I(\hat{s}) \mathrm{e}^{-2i\pi(ul+mv)} dldm \tag{2.10}$$

This is the case for telescopes with small fields of view where the sky can be considered flat. The intensity is recovered by taking the fourier transform of the complex visibilities.

2.2 Calibrating Interferometric Data

Astronomical radio signals are corrupted in a number of different ways. The radio interferometer measurement equation (RIME) is the fundamental tool for calibrating radio interferometric data (Hamaker et al. 1996). Here I will describe the RIME and its use. For a full review see Smirnov (2011a, b, c, d).

We can rewrite the correlator output for an uncorrupted signal as a 2×2 matrix which is equal to the product of the voltage of antenna one with the conjugate of voltage of antenna 2,

$$\mathbf{V}_{12} = 2\left\langle \begin{pmatrix} v_{1\mathrm{x}} \\ v_{1\mathrm{y}} \end{pmatrix} \begin{pmatrix} v_{2\mathrm{x}}^* & v_{2\mathrm{y}}^* \end{pmatrix} \right\rangle = 2\left\langle v_1 v_2^{\mathrm{H}} \right\rangle, \tag{2.11}$$

where v_{1x} and v_{1y} are the voltages from two orthogonal feeds on antenna one, v_{2x} and v_{2y} are the voltages from two orthogonal feeds on antenna two and H is the Hermitian transpose.

As the signal travels from the source to the antenna it is modified by different propagation effects like ionospheric Faraday rotation. Each of these effects can be represented by a 2×2 matrix and the product of these terms is the Jones matrix, **J**. The corrupted visibilities are then given by

$$\mathbf{V}_{12} = 2\left\langle \mathbf{J}_1 \mathbf{e} \left(\mathbf{J}_2 \mathbf{e}\right)^{\mathrm{H}} \right\rangle = 2\left\langle \mathbf{J}_1 \left(\mathbf{e}\mathbf{e}^{\mathrm{H}}\right) \mathbf{J}_2^{\mathrm{H}} \right\rangle = \mathbf{J}_1 \mathbf{B} \mathbf{J}_2^{\mathrm{H}}, \tag{2.12}$$

where $\mathbf{e}$ is the signal from the source and $\mathbf{B} = \left\langle \mathbf{e}\mathbf{e}^{\mathrm{H}} \right\rangle$ is the brightness matrix. An example Jones terms is given below

$$\mathbf{J} = \mathbf{BGDEPT}, \tag{2.13}$$

where **B** is the bandpass or frequency response of the antenna, **G** is the electronic gain response, **D** is called the D term and is the polarisation leakage between the orthogonal feeds, **E** represents the optical properties of the feed such as the shape of the primary beam, **P** is the parallactic angle term and takes into account the orientation of the feeds when projected onto the sky and **T** is the effects of the troposphere. Because matrices do not in general commute the order in which these effects are applied to the data is important.

2.2.1 *Ionospheric Effects*

As mentioned above **G** accounts for the complex electronic gains. In general direction independent ionospheric effects are also contained in the term **G**. The ionosphere is an ionised region of the upper atmosphere. The electron content in the ionosphere is highly variable, depending on both time of day and time of year. At low frequencies the ionosphere cause significant phase fluctuations. Assuming that the observing frequency, ν is much greater than the plasma frequency the ionospheric time delay, τ_{ion} is given by

$$\tau_{\mathrm{ion}} = \frac{40.3}{c\nu^2} \int n_{\mathrm{e}}(h) dh, \tag{2.14}$$

where $n_{\mathrm{e}}(h)$ is the electron density and $\int n_{\mathrm{e}}(h) dh$ is called the total electron content or (TEC).

At low frequencies, phase errors due to the ionosphere dominate and at higher frequencies phase errors due to the troposphere dominate. At approximately 1 GHz phase errors are minimised. In this thesis I will generally be working in the low frequency regime and calibration is largely focused with removing ionospheric effects.

In order to correct for ionospheric effects observations typically switch between the target source and a nearby compact bright calibrator source. It is assumed that the ionosphere in the direction of the calibrator is the same as the ionosphere in the direction of the target and that time variations in the ionosphere are longer than the time between calibrator scans. The complex gains **G** are solved for using a χ^2 minimisation algorithm on the calibrator source. These solutions are applied to the target source. Residual phase errors due to differences between the ionosphere along the target line of sight and calibrator line of sight can then be solved for using a process known as selfcal. During selfcal the target field is imaged. A model of the target field based on this image is then used for another round of complex gain solutions. This is done in an iterative fashion solving first for long timescale variations and moving to short timescale variations.

In the above calibration process it is assumed that the ionosphere is spatially constant over the full field of view. This is only appropriate for telescopes with a small field of view. However for telescopes such as the GMRT, LOFAR and the upcoming SKA, which have a very large field of view, direction dependent ionospheric effects must be taken into account. One direction dependent calibration pipeline is the Source Peeling and Atmospheric Modeling (SPAM) package (Intema 2014). The SPAM pipeline does an initial direction independent self cal. A model of the field is then subtracted and a number of compact bright sources are used to determine an ionospheric phase screen. This screen is then used to determine the complex gains which are then applied to the data. This is done twice to produce a final image of the sky with direction dependent effects removed. This pipeline has been successful used on GMRT and VLA data. Figure 2.1 shows a GMRT image before direction dependent calibration with SPAM and after.

LOFAR data has one of the largest fields of view available as well as observing at very low frequencies where direction dependent effects dominated the noise. The LOFAR field of view is so large that it is not necessary to observe an interleaved calibrator between the target scans. Instead a model of the field, generated from previous radio surveys (Smirnov and Noordam 2004) is used to do the initial phase calibration. LOFAR direction dependent calibration is not done using SPAM due to the difficulty the current implementation has in handling large LOFAR datasets. The DD calibration is done using FACTOR (van Weeren et al. 2016b). FACTOR calibration begins after an initial direction independent phase calibration. A model of the field is generated and subtracted from the visibilities to leave an 'empty' dataset. The sky is then divided into a number of facets. The centre point of each facet is located on a bright source or at the centre of a group of closely separated bright sources. It is assumed that the calibration solutions towards the bright source apply for the whole facet. The next step involves adding back a bright source or source group to the visibility data. The data is then phase rotated to the bright source. A number of selfcal cycles are performed in the direction of the bright source to obtain calibration solutions for the facet. The remaining sources in the facet are then added back to the visibility data. The calibration solutions are applied and the facet is images. The skymodel for that facet is then updated with the new model. This is repeated for every facet. With each facet the residual image becomes 'emptier'. This procedure has been tested and reaches the expected thermal noise of LOFAR data.

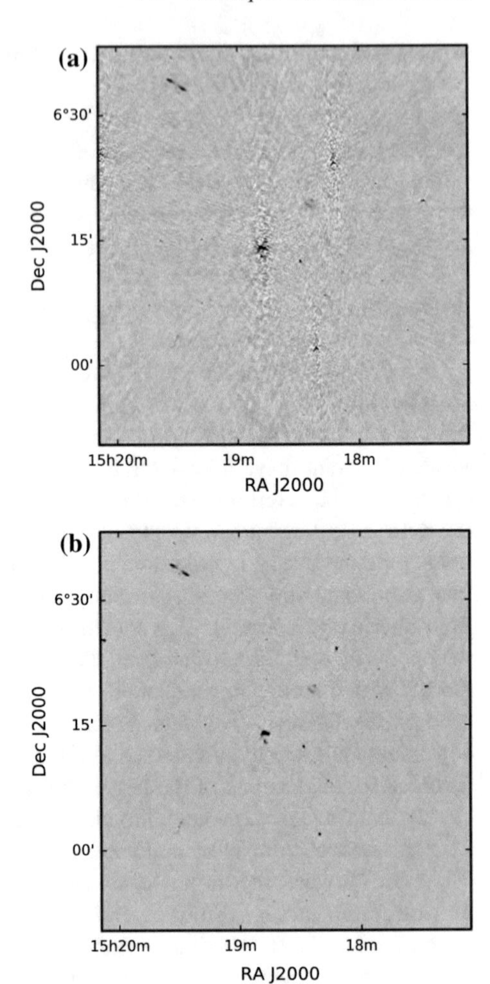

Fig. 2.1 **a** GMRT field before direction dependent calibration with SPAM. **b** GMRT field after direction dependent calibration with SPAM

2.3 Imaging Techniques

After radio data has been fully calibrated one effect is still present in the data and that is the effect of incomplete sampling of the visibilities. The interferometer does not measure the true visibilities but a sample of the visibilites, $\mathcal{V}_{\text{obs}}$ such that

$$\mathcal{V}_{\text{obs}} = W \times \mathcal{V}_{\text{true}}, \tag{2.15}$$

where W is 1 everywhere a visibility is measured and zero everywhere else. The Fourier transform of $\mathcal{V}_{\text{obs}}$ is

$$\text{FT}(\mathcal{V}_{\text{obs}}) = I * B_{\text{dirty}}, \tag{2.16}$$

where $B_{\text{dirty}} = \text{FT}(W)$ is known as the dirty beam. In order to reconstruct an image of the sky we need to deconvolve the dirty beam. This is done using the CLEAN algorithm (Högbom 1974). The basic steps of the clean algorithm are as follows.

1. Fourier transform the visibilities. This is known as the dirty image. The Fourier transform of the response to a point source is also computed. This is known as the dirty beam.
2. The peak in the dirty image is found. The dirty beam is subtracted from the position of this point source with an amplitude equal to the peak times some gain factor. The gain factor is less than one. This resulting image is now called the residual image. The position and amplitude of the subtracted component is added to a model image.
3. Step 2 is repeated until there is no obvious source structure left in the residual map.
4. The central peak of the dirty beam is fitted with a Gaussian. This Gaussian is the clean beam.
5. The model image is convolved with the clean beam and added to the residual map. This is the final image.

Figure 2.2a shows an image without deconvolution and Fig. 2.2b shows the same field but after deconvolution.

The most common implementation of the clean algorithm is the Cotton–Schwab clean (Schwab 1984). This method consists of major and minor cycles. In the minor cycles, the dirty beam is approximated to just the main beam and most dominant sidelobes. A number of points to subtract are found in the minor cycle. In the major cycle these points are subtracted from the data using the full dirty beam. The subtraction is done in the uv plane on the ungridded visibilities, i.e. it is the multiplication of the Fourier transform of the dirty beam and the Fourier transform of the points to be subtracted.

2.3.1 Multiscale Imaging

The early implementations of clean assumed the sky was a collection of point sources. Obviously this is not the case, and as the computing power available increased new algorithms were developed to take into account the extended structures. One such algorithm, which is widely used, is the multiscale clean algorithm (Wakker and Schwarz 1988; Cornwell 2008). Standard clean attempts to find the maximum in

$$I_{\text{R}}(n) = I_{\text{D}} - B * I_{\text{C}}(n-1), \tag{2.17}$$

where $I_{\text{R}}(n)$ is the residual map in the nth iteration of clean step 2, I_{D} is the dirty image, B is the dirty beam and $I_{\text{C}}(n-1)$ the model image after $n-1$ iterations of step 2. All components that contribute to $I_{\text{C}}(n-1)$ are delta components. Multiscale clean allows components to have different widths, α_1. Multiscale clean attempts to

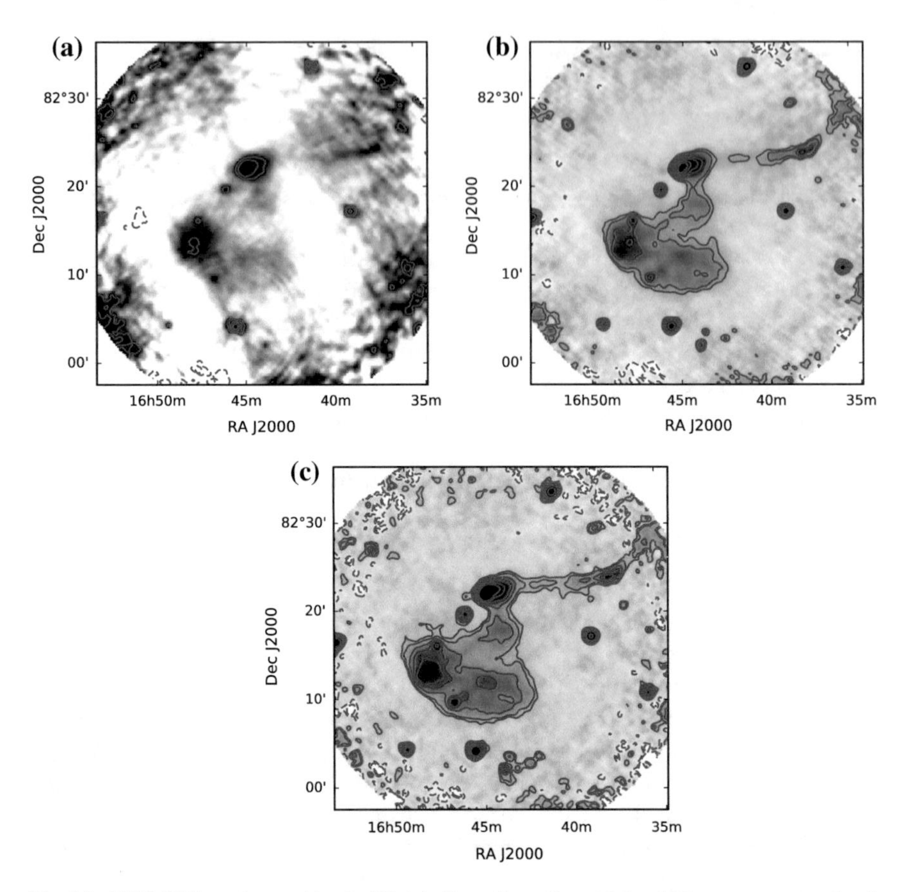

Fig. 2.2 NGC 6251 as observed by the VLA in D configuration at L band. Contours are at −5, −3, 3, 5, 10, 15, 20, 30, 40, 100 $\times\sigma_{\mathrm{rms}}$. **a** Dirty image, i.e. before any deconvolution. $\sigma_{\mathrm{rms}} = 3$ mJy **b** deconvolved image of NGC 6251 with no multiscsale clean $\sigma_{\mathrm{rms}} = 0.5$ mJy **c** deconvolved image of NGC 6251 with multiscale clean. $\sigma_{\mathrm{rms}} = 0.3$ mJy

find a maximum in

$$I(n) = I_{\mathrm{D}} - B * I_{\mathrm{M}}(n-1), \tag{2.18}$$

where $I_{\mathrm{M}} = \sum_{\mathrm{q}} I_{\mathrm{q}}(x - x_{\mathrm{q}}, y - y_{\mathrm{q}}, \alpha_{\mathrm{q}})$. The search for a peak is not just in x and y but also in component width. In practice only a finite range of component widths are searched. Figure 2.2 show a comparison between an image with and without multiscale clean. When no multiscale clean is used there is a large negative well around the extended emission. With multiscale clean this negative well is removed and the noise is reduced by a factor of 1.5.

2.3.2 W-Projection

In Sect. 2.1.2 I showed that for small field of views the w term can be neglected. However this assumption fails for many modern telescopes with wide fields of view. W-projection is an algorithm designed to image data taking the w term into account (Cornwell et al. 2008). The visibilities in Eq. 2.9 can be rewritten as

$$\mathcal{V}(u, v, w) = \int \frac{I(l, m)}{\sqrt{1 - l^2 - m^2}} G(l, m, w) e^{-2\pi i [ul + vm]} dl dm, \tag{2.19}$$

where $G(l, m, w) = e^{-2\pi i \left[w\left(\sqrt{1-l^2-m^2}-1\right)\right]}$. Thus the visibilities are given by

$$\mathcal{V}(u, v, w) = \tilde{G}(u, v, w) * V(u, v, w = 0), \tag{2.20}$$

where $\tilde{G}(u, v, w)$ is the Fourier transform of $G(l, m, w)$. We can therefore calculate the visibilities at any w from the convolution of the two dimensional visibilities at $w = 0$ with the know function $\tilde{G}(u, v, w)$. In order to image the visibilities of non coplanar baselines, each (u, v, w) point is projected onto the (u, v, w=0) plane. A 2-dimensional Fourier transform is used to produce the dirty image. Figure 2.3 shows a comparison of a wide field imaged with and without w-projection.

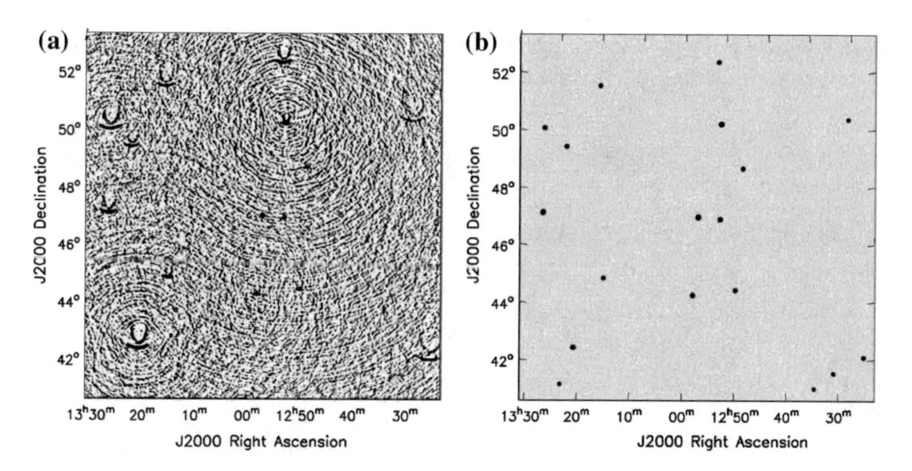

Fig. 2.3 Reproduced from Cornwell et al. (2008) **a** shows a image of simulated data without w-projection. **b** shows the same image but with w-projection

2.4 Rotation Measure Synthesis

Burn (1966) defines the Faraday dispersion function, $F(\phi)$, as $F(\phi) = E(\phi)P_{\text{true}}(\phi)$ where $P_{\text{true}}(\phi)$ is the intrinsic polarisation of radiation from Faraday depth ϕ and $E(\phi)$ is the fraction of radiation from Faraday depth ϕ. If the observed polarisation angle is given by $\chi = \chi_0 + \phi\lambda^2$ then the observed polarisation is given by

$$\begin{aligned} P(\lambda^2) &= \int_{-\infty}^{+\infty} E(\phi)P(\phi)e^{2i\phi\lambda^2}\mathrm{d}\phi, \\ P(\lambda^2) &= \int_{-\infty}^{+\infty} F(\phi)e^{2i\phi\lambda^2}\mathrm{d}\phi. \end{aligned} \tag{2.21}$$

Equation 2.21 has the form of a Fourier transform. However Eq. 2.21 only makes sense for $\lambda > 0$. Therefore in order to invert Eq. 2.21 we must make some assumptions about the behaviour of $P(\lambda^2)$ for $\lambda < 0$. One such assumption is that $P(\lambda^2)$ is hermitian, i.e. $P(-\lambda^2) = P^*(\lambda)$. Under this assumption $F(\phi)$ must be real.

A second complication arises from the fact that we do not have complete coverage of $\lambda^2 > 0$. (Brentjens and de Bruyn 2005) include this incomplete sampling so that Eq. 2.21 becomes

$$P_{\text{obs}}(\lambda^2) = W(\lambda^2)\int_{-\infty}^{+\infty} F(\phi)e^{2i\phi\lambda^2}\mathrm{d}\phi, \tag{2.22}$$

where $W(\lambda^2)$ is a weighting function which is 1 where there is data and zero everywhere else. Substituting $\lambda^2 = \pi u$ we get

$$P_{\text{obs}}(u\pi) = W(u\pi)\int_{-\infty}^{+\infty} F(\phi)e^{2i\pi\phi u}\mathrm{d}\phi, \tag{2.23}$$

They define the rotation measure transfer function (RMTF),

$$R(\phi) = \frac{\int_{-\infty}^{+\infty} W(u\pi)e^{2i\phi\lambda^2}\mathrm{du}}{\int_{-\infty}^{+\infty} W(u\pi)\mathrm{du}}, \tag{2.24}$$

so that

$$W(u\pi) = \left(\int_{-\infty}^{+\infty} W(u\pi)\mathrm{du}\right)\int_{-\infty}^{+\infty} R(\phi)e^{2i\pi\phi u}\mathrm{d}\phi, \tag{2.25}$$

and

$$P_{\text{obs}}(u\pi) = \left(\int_{-\infty}^{+\infty} W(u\pi)\mathrm{du}\right)\int_{-\infty}^{+\infty} R(\phi)e^{2i\pi\phi u}\mathrm{d}\phi \times \int_{-\infty}^{+\infty} F(\phi)e^{2i\pi\phi u}\mathrm{d}\phi. \tag{2.26}$$

Taking the Fourier transform of Eq. 2.26

$$\begin{aligned} F(\phi) * R(\phi) &= \frac{\int_{-\infty}^{+\infty} P_{\text{obs}}(u\pi)e^{2i\pi\phi u}\text{du}}{\int_{-\infty}^{+\infty} W(u\pi)\text{du}}, \\ F(\phi) * R(\phi) &= K \int_{-\infty}^{+\infty} P_{\text{obs}}(\lambda^2)e^{2i\phi\lambda^2}\text{d}\lambda^2. \end{aligned} \tag{2.27}$$

Taking the Fourier transform of the observed polarisation as a function of wavelength squared yields the Faraday spectrum convolved with the rotation measure transfer function.

Figure 2.4 shows an example of what a Faraday spectrum could look like for different emitting regions along the line of sight. Faraday thick regions, i.e. regions where $\lambda\Delta\phi \gg 1$, are extended regions in the Faraday spectrum and Faraday thin components, i.e. regions where $\lambda\Delta\phi \ll 1$, appear as narrow peaks. With sufficiently good resolution in Faraday space individual components along the line of sight can be resolved in the Faraday spectrum.

The maximum observable Faraday depth, $\phi_{\text{max-depth}}$, the resolution in Faraday space, $\delta\phi$, and the largest scale in Faraday space that can be detected, $\phi_{\text{max-scale}}$, are given by (Brentjens and de Bruyn 2005)

$$\|\phi_{\text{max-depth}}\| \approx \frac{\sqrt{3}}{\delta\lambda^2} \tag{2.28a}$$

$$\delta\phi \approx \frac{2\sqrt{3}}{\Delta\lambda^2} \tag{2.28b}$$

$$\phi_{\text{max-scale}} \approx \frac{\pi}{\lambda^2_{\text{min}}}, \tag{2.28c}$$

where $\delta\lambda^2$ is the channel width squared, $\Delta\lambda^2$ is the width of the λ^2 distribution and λ^2_{min} is the minimum wavelength squared. Figure 2.5 shows the RMSF for two diiferent telescopes, the VLA at 3 GHz and LOFAR at 150 MHz. The resolution of LOFAR is significantly better than that of the VLA due to LOFAR's large bandwidth in λ^2. However the VLA has much better sensitivity to extended structures in Faraday space due to a lower λ^2_{min}.

2.5 QU Fitting

Another method which takes advantage of wide bandwidth telescopes is the QU fitting method. QUfitting attempts to fit models of polarised intensity to $Q(\lambda^2)$ and $U(\lambda^2)$. Typical models used are those for a Faraday thin screen and Faraday thick screen. The polarised signal from a Faraday thin screen is given by

$$P(\lambda^2) = p \exp\left[2i\left(\chi + \phi\lambda^2\right)\right]. \tag{2.29}$$

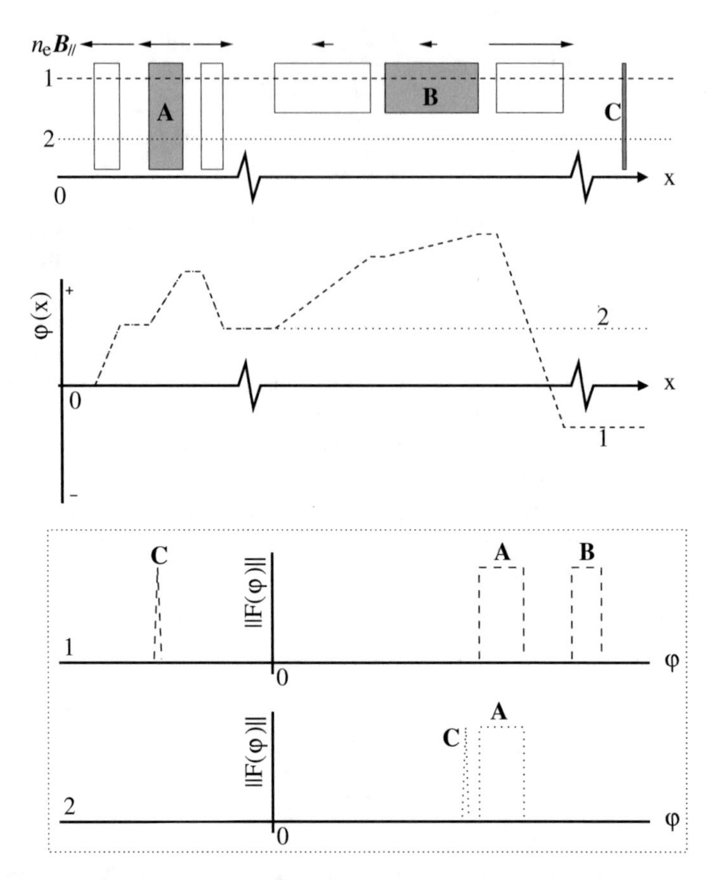

Fig. 2.4 The top panel shows two lines of sight which pass through emitting regions (grey box), regions which are both emitting and rotating (grey boxes with arrows) and rotating regions (white boxes with arrows). The arrows represent the strength and direction of the magnetic field in different regions. The middle panel plots the observed Faraday depth as a function of position along the line of sight. The bottom panel shows the Faraday spectrum for the two lines of sight. Regions A and B, which are both emiiting rotating regions, are extended in Faraday space while region C, which is only emitting, apears as a narrow peak in the Faraday spectrum. Figure 2.2 from Brentjens and de Bruyn (2005)

where $P(\lambda^2)$ is the observed polarised intensity, p is the initial polarised intensity, χ is the polarisation angle. The polarised signal from a Faraday thick slab is given by

$$P(\lambda^2) = p\frac{\sin(\phi_s\lambda^2)}{\phi_s\lambda^2}\exp\left[2i\left(\chi_0 + \phi_0\lambda^2 + \frac{1}{2}\phi_0\lambda^2\right)\right], \qquad (2.30)$$

where ϕ_s is the extent of the slab in Faraday space, ϕ_0 is the front edge of the screen and χ_0 is a constant. Other more complicated models could also be fit to data for example a Faraday thick slab with a varying polarisation angle or Gaussian regions. Sun et al.

(2015) compare the ability QU fitting algorithms and RM synthesis algorithms to fit both simple and complex Faraday spectra. They find that both types of algorithms work well for spectra containing a single Faraday thin component. Only QU fitting was capable of fitting spectra containing two Faraday thin components. It was found that neither algorithm was particularly good at reproducing Faraday thick spectra due in part to the narrow bandwidth used in the simulations. QU fitting is therefore an important tool for the analysis of complex Faraday structures.

QU fitting is already being used to produce interesting science. O'Sullivan et al. (2015) examine the polarisation properties of a large sample of AGN using QU fitting. They show that not only does the density of the host environment play a role in the accretion state of AGN but that the magnetised properties of that environment is important in the formation of AGN jets.

2.6 Instruments

In this thesis I used interferometric data from four different instruments, the GMRT, LOFAR, the VLA, and KAT-7. Below I describe each of these instruments and how they differ from each other.

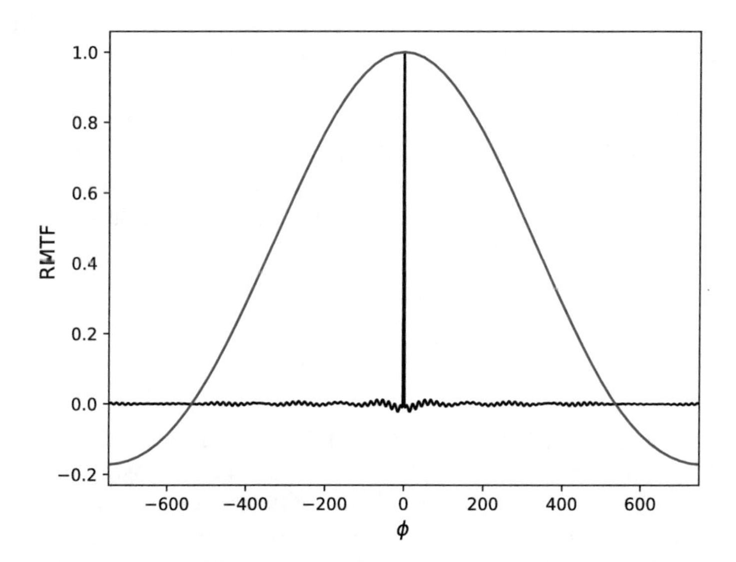

Fig. 2.5 RMTF for LOFAR (black) and the VLA (blue)

2.6.1 GMRT

The Giant Meter-Wave Radio Telescope (GMRT) (Swarup et al. 1991) is an array of 30 antennas with diameters of 45 m in India. 14 antennas make up the central core of the array while the remaining 16 antennas are arranged around the core to form a Y-shape. This distribution of antennas affords good *uv* coverage on both the short and long baselines. The GMRT is capable of observing at a range of frequencies from 150 to 1280 MHz with a maximum bandwidth of 33 MHz. At 610 MHz the maximum resolution is about 5 arcsec and the half power point of the primary beam is 43 arcmin.

2.6.2 LOFAR

The LOw Frequency ARray (LOFAR) (van Haarlem et al. 2013) is an array of dipole antennas which are arranged into groups of "stations". LOFAR is capable of observing from 110 to 250 MHz using the high band antenna (HBA) and 30–90 MHz using the low band antenna (LBA).

LOFAR consists of 24 "core" stations, 14 "remote" stations in the Netherlands and 12 international stations in Germany, UK, France, Sweden, Poland and Ireland. Figure 2.6 shows the layout of the core, remote and international stations. The core and remote stations will be refered to as the Dutch array for the remainder of this section.

Core stations have a maximum baseline of 3.5 km, remote stations have a maximum baseline of 121 Km and including the international stations increases the maximum baseline to $\sim$1900 km. The angular resolution of the Dutch array is $\sim$8 arcsec for the LBA and $\sim$5 arcsec for the HBA. The Dutch array has a field of view of 20 deg^2 for the LBA and 7 deg^2 for the HBA.

2.6.3 VLA

The Karl G. Jansky Very Large Array (VLA) (Thompson et al. 1980; Napier et al. 1983; Perley et al. 2011) is an array of twenty-seven antennas each with a diameter of 25m. The telescopes can be arranged in four different configurations each with a basic Y-shape. The configurations are named A, B, C and D configuration with A being the least compact configuration and D being the most compact configuration. The VLA can observe from 54 to 50 GHz with continuous coverage from 1 to 50 GHz.

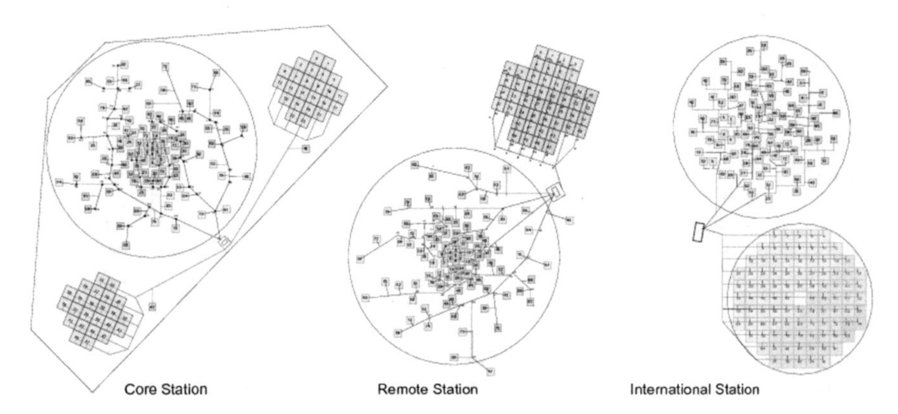

Fig. 2.6 Configuration of LOFAR stations. The circles mark the LBA antennas and the square tiles show the HBA tile layouts. Figure 4 from van Haarlem et al. (2013)

2.6.4 KAT-7

The Karoo Array Telescope (KAT-7)[1] is an array of 7 antennas with diameters of 12 m in South Africa with a longest baseline of 200 m. KAT-7 can observe at central frequencies of 1382 MHz and 1826 MHz with a bandwidth of 256 MHz. The maximum resolution of KAT-7 is 4 arcmin at 1382 MHz and 3 arcmin at 1826 MHz.

References

Nan R, Li D, Jin C, Wang Q, Zhu L, Zhu W, Zhang H, Yue Y, Qian L (2011) The five-hundred aperture spherical radio telescope (fast) project. Int J Mod Phys D 20:989–1024. https://doi.org/10.1142/S0218271811019335. arXiv:1105.3794

Thompson AR, Moran JM, Swenson GW Jr (2017) Interferometry and synthesis in radio astronomy, 3rd edn. https://doi.org/10.1007/978-3-319-44431-4

Hamaker JP, Bregman JD, Sault RJ (1996) Understanding radio polarimetry. I. Mathematical foundations. A&AS 117:137–147

Smirnov OM (2011a) Revisiting the radio interferometer measurement equation. I. A full-sky Jones formalism. A&A 527:A106. https://doi.org/10.1051/0004-6361/201016082. arXiv:1101.1764

Smirnov OM (2011b) Revisiting the radio interferometer measurement equation. II. Calibration and direction-dependent effects. A&A 527:A107. https://doi.org/10.1051/0004-6361/201116434. arXiv:1101.1765

Smirnov OM (2011c) Revisiting the radio interferometer measurement equation. III. Addressing direction-dependent effects in 21 cm WSRT observations of 3C 147. A&A 527:A108. https://doi.org/10.1051/0004-6361/201116435. arXiv:1101.1768

Smirnov OM (2011d) Revisiting the radio interferometer measurement equation. IV. A generalized tensor formalism. A&A 531:A159. https://doi.org/10.1051/0004-6361/201116764. arXiv:1106.0579

[1]For more information see www.ska.ac.za/science-engineering/kat-7/.

Intema HT (2014) SPAM: a data reduction recipe for high-resolution, low-frequency radio-interferometric observations. In: Astronomical Society of India Conference Series, Astronomical Society of India Conference Series, vol 13. arXiv:1402.4889

Smirnov OM, Noordam JE (2004) The LOFAR global sky model: some design challenges. In: Ochsenbein F, Allen MG, Egret D (eds) Astronomical data analysis software and systems (ADASS) XIII, Astronomical Society of the Pacific Conference Series, vol 314, p 18

van Weeren RJ, Williams WL, Hardcastle MJ, Shimwell TW, Rafferty DA, Sabater J, Heald G, Sridhar SS, Dijkema TJ, Brunetti G, Brüggen M, Andrade-Santos F, Ogrean GA, Röttgering HJA, Dawson WA, Forman WR, de Gasperin F, Jones C, Miley GK, Rudnick L, Sarazin CL, Bonafede A, Best PN, Bîrzan L, Cassano R, Chyy KT, Croston JH, Ensslin T, Ferrari C, Hoeft M, Horellou C, Jarvis MJ, Kraft RP, Mevius M, Intema HT, Murray SS, Orrú E, Pizzo R, Simionescu A, Stroe A, van der Tol S, White GJ (2016b) LOFAR facet calibration. ApJS 223:2. https://doi.org/10.3847/0067-0049/223/1/2. arXiv:1601.05422

Högbom JA (1974) Aperture synthesis with a non-regular distribution of interferometer baselines. A&AS 15:417

Schwab FR (1984) Relaxing the isoplanatism assumption in self-calibration; Applications to low-frequency radio interferometry. AJ 89:1076–1081. https://doi.org/10.1086/113605

Wakker BP, Schwarz UJ (1988) The Multi-Resolution CLEAN and its application to the short-spacing problem in interferometry. A&A 200:312–322

Cornwell TJ (2008) Multiscale CLEAN deconvolution of radio synthesis images. IEEE J Sel Top Signal Process 2:793–801. https://doi.org/10.1109/JSTSP.2008.2006388

Cornwell TJ, Golap K, Bhatnagar S (2008) The noncoplanar baselines effect in radio interferometry: the w-projection algorithm. IEEE J Sel Top Signal Process 2:647–657. https://doi.org/10.1109/JSTSP.2008.2005290. arXiv:0807.4161

Burn BJ (1966) On the depolarization of discrete radio sources by Faraday dispersion. MNRAS 133:67. https://doi.org/10.1093/mnras/133.1.67

Brentjens MA, de Bruyn AG (2005) Faraday rotation measure synthesis. A&A 441:1217–1228. https://doi.org/10.1051/0004-6361:20052990. arXiv:astro-ph/0507349

Sun XH, Rudnick L, Akahori T, Anderson CS, Bell MR, Bray JD, Farnes JS, Ideguchi S, Kumazaki K, O'Brien T, O'Sullivan SP, Scaife AMM, Stepanov R, Stil J, Takahashi K, van Weeren RJ, Wolleben M (2015) Comparison of algorithms for determination of rotation measure and faraday structure. I. 1100–1400 MHz. AJ 149:60. https://doi.org/10.1088/0004-6256/149/2/60. arXiv:1409.4151

O'Sullivan SP, Gaensler BM, Lara-López MA, van Velzen S, Banfield JK, Farnes JS (2015) The magnetic field and polarization properties of radio galaxies in different accretion states. ApJ 806:83. https://doi.org/10.1088/0004-637X/806/1/83. arXiv:1504.06679

Swarup G, Ananthakrishnan S, Kapahi VK, Rao AP, Subrahmanya CR, Kulkarni VK (1991) The giant metre-wave radio telescope. Curr Sci 60(2):95

van Haarlem MP, Wise MW, Gunst AW, Heald G, McKean JP, Hessels JWT, de Bruyn AG, Nijboer R, Swinbank J, Fallows R, Brentjens M, Nelles A, Beck R, Falcke H, Fender R, Hörandel J, Koopmans LVE, Mann G, Miley G, Röttgering H, Stappers BW, Wijers RAMJ, Zaroubi S, van den Akker M, Alexov A, Anderson J, Anderson K, van Ardenne A, Arts M, Asgekar A, Avruch IM, Batejat F, Bähren L, Bell ME, Bell MR, van Bemmel I, Bennema P, Bentum MJ, Bernardi G, Best P, Bîrzan L, Bonafede A, Boonstra AJ, Braun R, Bregman J, Breitling F, van de Brink RH, Broderick J, Broekema PC, Brouw WN, Brüggen M, Butcher HR, van Cappellen W, Ciardi B, Coenen T, Conway J, Coolen A, Corstanje A, Damstra S, Davies O, Deller AT, Dettmar RJ, van Diepen G, Dijkstra K, Donker P, Doorduin A, Dromer J, Drost M, van Duin A, Eislöffel J, van Enst J, Ferrari C, Frieswijk W, Gankema H, Garrett MA, de Gasperin F, Gerbers M, de Geus E, Grießmeier JM, Grit T, Gruppen P, Hamaker JP, Hassall T, Hoeft M, Holties HA, Horneffer A, van der Horst A, van Houwelingen A, Huijgen A, Iacobelli M, Intema H, Jackson N, Jelic V, de Jong A, Juette E, Kant D, Karastergiou A, Koers A, Kollen H, Kondratiev VI, Kooistra E, Koopman Y, Koster A, Kuniyoshi M, Kramer M, Kuper G, Lambropoulos P, Law C, van Leeuwen J, Lemaitre J, Loose M, Maat P, Macario G, Markoff S, Masters J, McFadden

RA, McKay-Bukowski D, Meijering H, Meulman H, Mevius M, Middelberg E, Millenaar R, Miller-Jones JCA, Mohan RN, Mol JD, Morawietz J, Morganti R, Mulcahy DD, Mulder E, Munk H, Nieuwenhuis L, van Nieuwpoort R, Noordam JE, Norden M, Noutsos A, Offringa AR, Olofsson H, Omar A, Orrú E, Overeem R, Paas H, Pandey-Pommier M, Pandey VN, Pizzo R, Polatidis A, Rafferty D, Rawlings S, Reich W, de Reijer JP, Reitsma J, Renting GA, Riemers P, Rol E, Romein JW, Roosjen J, Ruiter M, Scaife A, van der Schaaf K, Scheers B, Schellart P, Schoenmakers A, Schoonderbeek G, Serylak M, Shulevski A, Sluman J, Smirnov O, Sobey C, Spreeuw H, Steinmetz M, Sterks CGM, Stiepel HJ, Stuurwold K, Tagger M, Tang Y, Tasse C, Thomas I, Thoudam S, Toribio MC, van der Tol B, Usov O, van Veelen M, van der Veen AJ, ter Veen S, Verbiest JPW, Vermeulen R, Vermaas N, Vocks C, Vogt C, de Vos M, van der Wal E, van Weeren R, Weggemans H, Weltevrede P, White S, Wijnholds SJ, Wilhelmsson T, Wucknitz O, Yatawatta S, Zarka P, Zensus A, van Zwieten J (2013) LOFAR: the low-frequency array. A&A 556:A2. https://doi.org/10.1051/0004-6361/201220873. arXiv:1305.3550

Thompson AR, Clark BG, Wade CM, Napier PJ (1980) The very large array. ApJS 44:151–167. https://doi.org/10.1086/190688

Napier PJ, Thompson AR, Ekers RD (1983) The very large array–design and performance of a modern synthesis radio telescope. IEEE Proc 71:1295–1320

Perley RA, Chandler CJ, Butler BJ, Wrobel JM (2011) The expanded very large array: a new telescope for new science. ApJ 739:L1. https://doi.org/10.1088/2041-8205/739/1/L1. arXiv:1106.0532

Chapter 3
A Newly-Discovered Radio Halo in Merging Cluster MACS J2243.3-093

The work in this chapter was published in Cantwell et al. (2016). I personally did all work presented in the paper except the observation and data reducation of the KAT-7 data presented in Sect. 3.2.1 which were carried out by our collaborator N. Oozeer.

3.1 Introduction

MACS J2243.3-0935 is a massive galaxy cluster at a redshift of $z = 0.447$ at the center of the super cluster SCL2243-0935 (Schirmer et al. 2011). Table 3.1 lists some important properties of MACS J2243.3-0935 determined from previous studies across a range of wavelengths (Ebeling et al. 2010; Planck Collaboration et al. 2014; Mantz et al. 2010; Mann and Ebeling 2012; Wen and Han 2013).

The dynamical state of this cluster has been examined using a variety of techniques. Mann and Ebeling (2012) use the X-ray morphology and the offset between the brightest cluster galaxy (BCG) and X-ray peak/centroid to characterise the dynamical state of clusters. They find that the merger axis suggested by the highly elongated X-ray emission in MACS J2243.3-0935 is mis-aligned with the merger axis suggested by the two main galaxy concentrations. Mann and Ebeling (2012) also report a separation between the BCG and the X-ray peak of 125 ± 6 kpc and a centroid shift, w, of 156 ± 4 kpc.

Wen and Han (2013) calculate the relaxation parameter , Γ, of MACS J2243.3-0935 to be -1.53 ± 0.07. The highly negative relaxation parameter of MACS J2243.3-0935, as well as its high luminosity, were the primary reasons for selecting this cluster for study in this work.

MACS J2243.3-0935 has also been detected by *Planck* as PSZ2 G056.93-55.08 (Planck Collaboration et al. 2015). From these data, the total mass measured from the SZ effect for this cluster is 1.007×10^{15} $M_{\odot}$ and the cluster has an integrated Compton-y parameter, $Y = 16.3 \pm 1.7$ arcsec^2. In the radio, the cluster field was

T. Cantwell, *Low Frequency Radio Observations of Galaxy Clusters and Groups*, Springer Theses, https://doi.org/10.1007/978-3-319-97976-2_3

Table 3.1 Properties of MACS J2243.3-0935

Property	Value	Reference
z	0.447	Ebeling et al. (2010)
M_{500} ($\times 10^{14} M_{\odot}$)	10.07 ± 0.58	Planck Collaboration et al. (2014)
Y_{500} (Mpc^2)	16.3 ± 1.7	Planck Collaboration et al. (2014)
$L_{x,500}$ (10^{44}erg s^{-1})	11.56 ± 0.67	Mantz et al. (2010)
T (keV)	8.24 ± 0.92	Mantz et al. (2010)
w (kpc)	156 ± 4	Mann and Ebeling (2012)
Γ	-1.53 ± 0.07	Wen and Han (2013)
Virial radius (Mpc)	$2.13^{+0.18}_{-0.12}$	Schirmer et al. (2011)

L_x is core excised from 0.1–2.4 kev

observed by the NRAO VLA Sky Survey (NVSS) (Condon et al. 1998a) and the Faint Images of the Radio Sky at Twenty-cm survey (FIRST) (Becker et al. 1995) however the field is not covered by the Sydney University Molonglo Sky Survey (SUMSS) (Bock et al. 1999) or the Westerbork Northern Sky Survey (WENSS) (Rengelink et al. 1997).

In this chapter I present new observations of MACS J2243.3-0935 at 1.4 GHz using the Karoo Array Telescope[1] (KAT-7) and at 610 MHz with the Giant Meter Wave Telescope (GMRT) (Swarup et al. 1991). In Sect. 3.2 I describe the observations and data reduction. In Sect. 3.3 I present the results which are then discussed in Sect. 3.4 before making my concluding remarks in Sect. 3.5.

At a redshift of 0.447, 1 arcsec corresponds to a physical scale of 5.74 kpc (Wright 2006).

3.2 Observations and Data Reduction

MACS J2243.3-0935 was observed at 1.8GHz by KAT-7 and at 610 MHz with the GMRT. See Sect. 2.6 for more details on these instruments.

A summary of the observational details can be found in Table 3.2.

3.2.1 KAT-7

KAT-7 was used to observe MACS J2243.3-0935 on the 2012-09-08. All seven antennas were used for this observation at a frequency of 1822 MHz with a bandwidth of 400 MHz. Due to analog filters in the IF and baseband system only the central 256

[1] For more information see http://public.ska.ac.za/kat-7.

Table 3.2 Observation details of MACS J2243.3-0935

Telescope	GMRT	KAT-7
Date	20 Jun 2014	7 Sep 2012
Frequency (MHz)	610	1826
Time on Target (hrs)	5.6	7.5
Usable Time (hrs)	5.6	7.5
Bandwidth (MHz)	33	400
Usable Bandwidth (MHz)	29	256
No. Channels	256	600
No. Averaged Channels	28	9
% flagged	33%	18.5%
Sensitivity	40 μJy	500 μJy
Angular Resolution	~5 arcsec	~3 arcmin
FOV	43 arcmin	~60 arcmin

MHz of the bandwidth is useable. A first round of flagging was carried out by the automatic flagging routine (developed in-house) to remove known radio frequency interference (RFI). The data were further flagged inside CASA to remove other low level RFI. PKS 1934-638 was used as the primary calibrator and PKS2243-123 as the phase calibrator. The flux calibrator was observed every 2 h for 2min while the phase calibrator was observed every 15min for 3 min. Flux densities were tied to the Perley-Butler-2010 flux density scale (Perley and Butler 2013). Standard data flagging and calibration was carried out in CASA4.3. Three rounds of phase only selfcal were performed. The data were then imaged using the multifrequency, multiscale clean task in CASA with a Briggs weighting robust parameter of 0. The resulting image has an rms of 0.5 mJy. Figure 3.1 shows the full field KAT-7 image of MACS J2243.3-0935 while Fig. 3.2 shows the central cluster region.

3.2.2 *GMRT*

MACS J2243.3-0935 was observed by the GMRT at 610 MHz with a bandwidth of 33 MHz on 20th of June 2014.[2] The primary calibrator, 3C48, was observed for fifteen minutes at the start of the observations and 3C468.1 was observed at the end of the observation in order to test the flux calibration. The phase calibrator J2225-049 was observed for 5 min, every 20 min. Data reduction and calibration for GMRT data at 610 MHz were carried out in CASA (McMullin et al. 2007). The calibration process

[2]MACS J2243.3-0935 was also observed by the GMRT for 7 h on 25th October 2010 at 610 Mz and 235 MHz. These observations were taken before upgrades began on the GMRT and are of lower quality than the new observations and are significantly contaminated with RFI. Including these data does not improve the image quality.

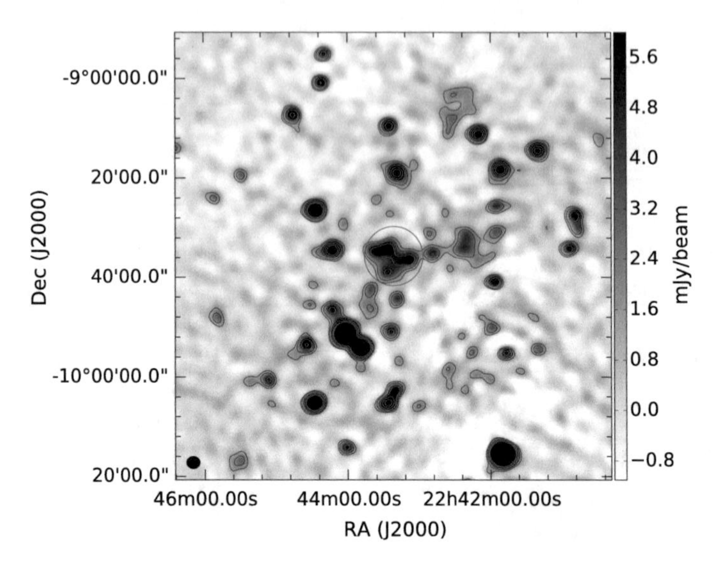

Fig. 3.1 Greyscale plot showing the KAT-7 FOV image of MACS J2243.3-0935 with KAT-7 contours overlaid in red. Contours are at 3, 5, 10, 15, 20 × $\sigma_{\rm rms}$ where $\sigma_{\rm rms} = 500$ μJy/beam. The resolution is 160.10 × 144.99 arcsec. The blue circle marks the virial radius of MACS J2243.3-0935. The virial radius is $2.13^{+0.18}_{-0.12}$ Mpc or 370 arcsec

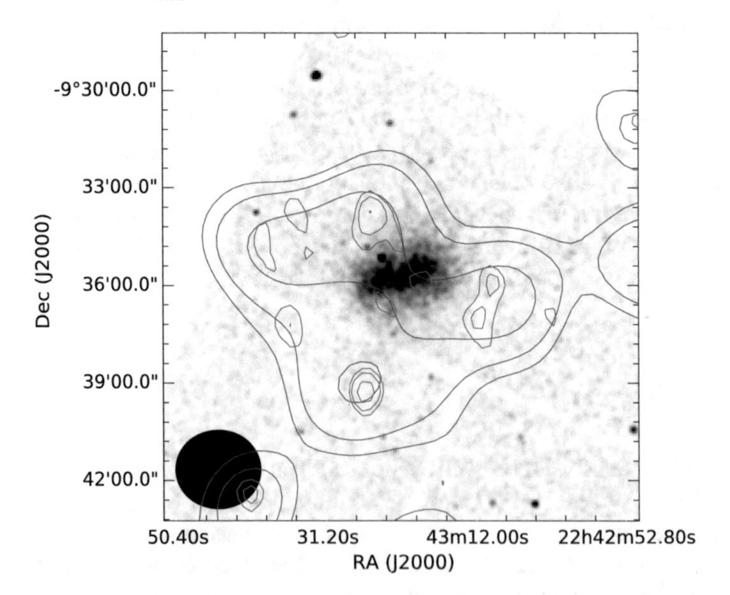

Fig. 3.2 Greyscale plot showing X-ray image of MACS J2243.3-0935 in the Chandra ACIS 0.5−7 keV band. The image has been smoothed with a Gaussian kernel with $\sigma = 3$ pixels. KAT-7 contours are overlaid in red while NVSS contours are overlaid in blue. Contours are at 3, 5, 10, 15, 20 × $\sigma_{\rm rms}$ where $\sigma_{\rm rms} = 500$ μJy/beam for KAT-7 and $\sigma_{\rm rms} = 400$ μJy/beam for NVSS. The resolution of the KAT-7 image is 160.10 × 144.99 arcsec while the resolution of the NVSS image is 45 × 45 arcsec

followed that described in De Gasperin et al. (2014). Calibration and flagging were performed in an iterative fashion. Data were phase, amplitude and bandpass calibrated then flagged, in the first round, using first the rflag mode in the CASA task flagdata and, in the second round, with AOFlagger (Offringa et al. 2012). After flagging with AOFlagger, the data were averaged and a final round of calibration and flagging with AOFlagger was performed. Five rounds of phase only selfcal were carried out. After calibration and selfcal, approximately 33% of the data were flagged. Flux densities were tied to the Perley-Butler-2010 flux density scale (Perley and Butler 2013). The flux density measured from 3C468.1 was 12.5 ± 0.9 Jy, which agrees within error with the literature value of 12.7 Jy (Helmboldt et al. 2008; Pauliny-Toth et al. 1966). The data were then imaged using the multifrequency, multiscale clean task in CASA.

To subtract the point source population from the centre of the cluster, data from baselines longer than 4 kλ, which corresponds to an angular scale of 50 arcsec, were imaged with a Briggs robust parameter of 0. This image is shown in Fig. 3.3a and has an rms of 40 μJy/beam and a resolution of 4.84×4.15 arcsec. Figure 3.3b shows the

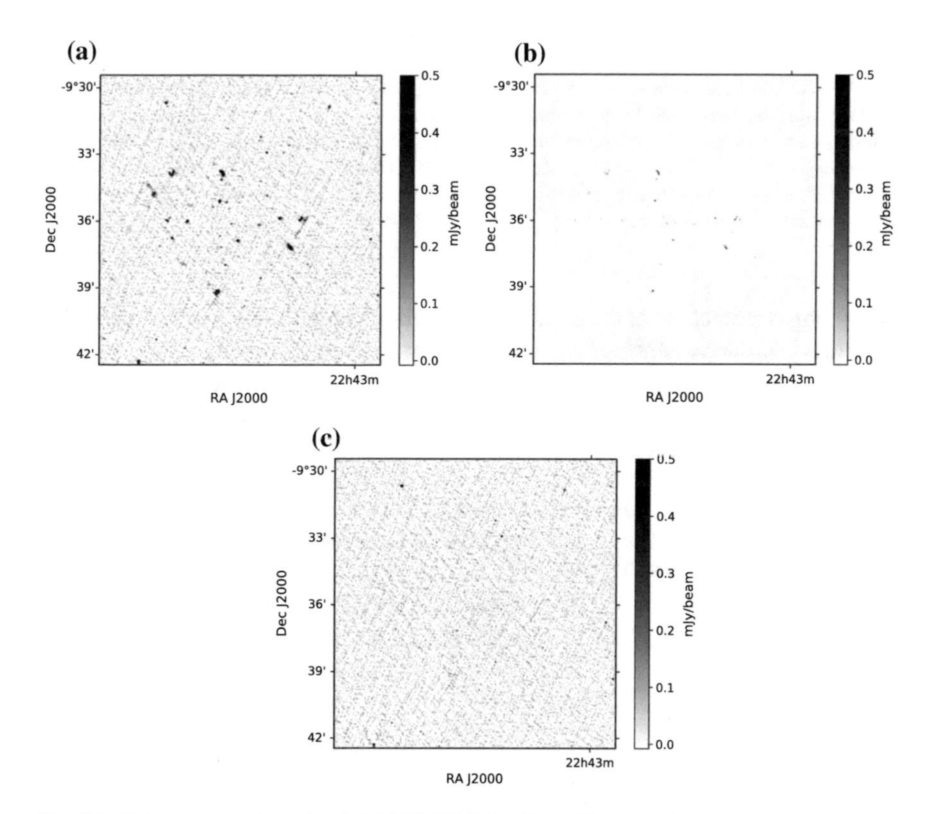

Fig. 3.3 Point source subtraction in MACS J2243.3-0935. **a** Greyscale image shows the robust 0 high resolution GMRT data used for the point source subtraction. **b** Greyscale image shows the clean component model of the point sources in the centre of the cluster **c** Greyscale image shows the robust 0 high resolution GMRT data after point source subtraction

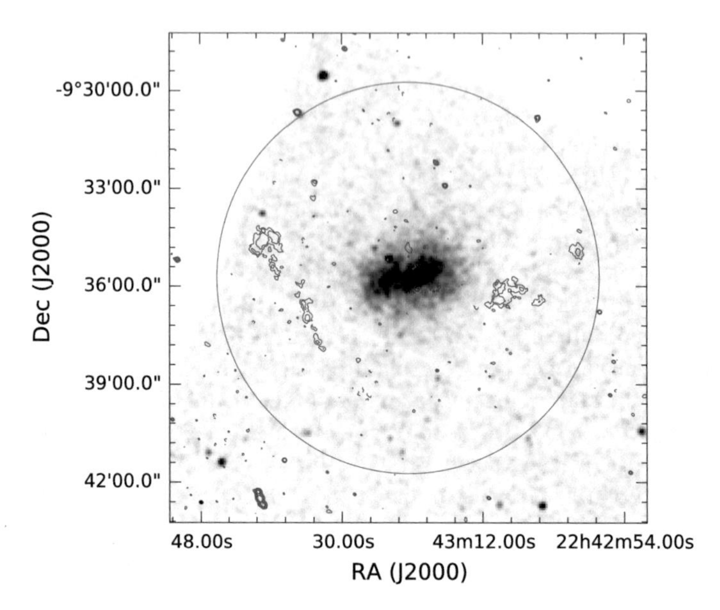

Fig. 3.4 Greyscale plot showing X-ray image of MACS J2243.3-0935 in the Chandra ACIS 0.5–7 keV band. The image has been smoothed with a Gaussian kernel with $\sigma = 3$ pixels. The red contours show the naturally weighted GMRT image. Contours are at $-3, 3, 5, 10, 15, 20 \times \sigma_{\mathrm{rms}}$ where $\sigma_{\mathrm{rms}} = 45$ μJy/beam. The resolution of the GMRT image is 7.44 × 6.06 arcsec. Radio point sources have been subtracted from this image. The blue circle marks the virial radius of MACS J2243.3-0935. The virial radius is $2.13^{+0.18}_{-0.12} Mpc$ or 370 arcsec

clean component model of the point sources in the clusters. These clean components were then Fourier transformed and subtracted from the uv data using the CASA tasks ft and uvsub. In order to check the quality of the point source subtraction the data were reimaged with a uvcut of 4 kλ and a Briggs robust paramter of 0. Figure 3.3c shows the high reslution image after subtraction. The bright sources at the centre of the cluster are no longer present.

Two images were then made using the point source subtracted data: the first was naturally weighted with no uvtaper and the second was naturally weighted with a uvtaper of 5 kλ by 4 kλ. The naturally weighted image has a resolution of 7.44 × 6.06 arcsec and an rms of 45 μJy/beam and is shown in Fig. 3.4. The tapered image with a resolution of 44.89 × 33.70 arcsec and an rms of 200 μJy/beam is shown in Fig. 3.5.

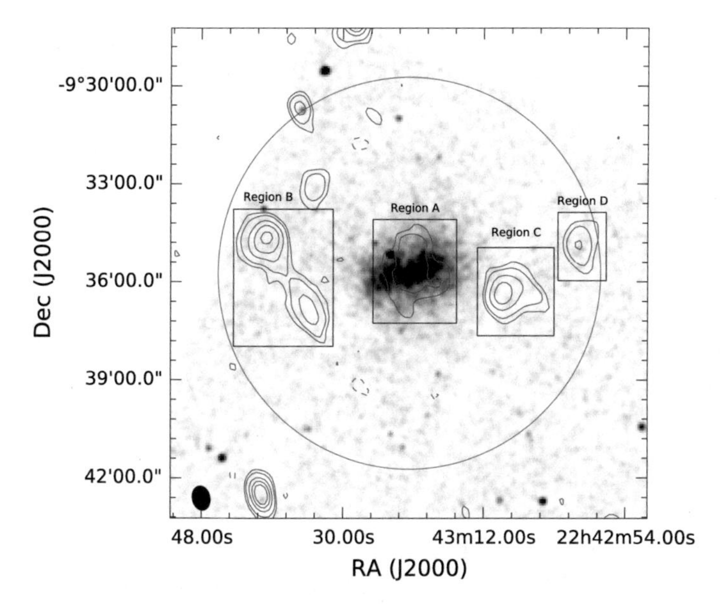

Fig. 3.5 Greyscale plot showing X-ray image of MACS J2243.3-0935 in the Chandra ACIS 0.5–7 keV band. The image has been smoothed with a Gaussian kernel with $\sigma = 3$ pixels. The red contours show the naturally weighted GMRT image with a uvtaper of 5×4 kλ. Contours are at $-3, 3, 5, 10, 15, 20 \times \sigma_{\rm rms}$ where $\sigma_{\rm rms} = 200$ μJy/beam. The resolution of the GMRT image is 44.89×33.70 arcsec. Radio point sources have been subtracted from this image. The blue circle marks the virial radius of MACS J2243.3-0935. The virial radius is $2.13^{+0.18}_{-0.12} Mpc$ or 370 arcsec. Different regions of diffuse radio emission are marked with black rectangles and labelled

3.3 Results

3.3.1 MACS J2243.3-0935 at >1 GHz

Figure 3.1 shows the KAT-7 image of MACS J2243.3-0935 at 1822 MHz. The resolution of the image is 160.10×144.99 arcsec. However, the fidelity of the KAT-7 image is limited by the presence of a bright complex source to the south west of the field of view (FOV). A large region of diffuse emission can be seen towards the cluster center with a largest linear scale (LLS) of 3.9 Mpc. This diffuse emission is detected at a 5σ level. The flux density of the cluster emission detected by KAT-7 , measured within the 3σ level, is 40 ± 6 mJy. Comparison with NVSS in Fig. 3.2 shows that there are at least two unresolved point sources in this region. Given the low resolution of these data it is not possible to characterise the emission in the KAT-7 image further than to note that diffuse emission appears to be present in addition to these compact sources.

Table 3.3 Properties of the diffuse emission in regions A, B, C and D. Column 1 is the region name, column 2 is the integrated flux density at 610 MHz, column 3 is the k-corrected integrated flux density as 610 MHz, column 4 is the surface brightness and column 5 is the k-corrected power at 610 MHz

Region	S_{610MHz} (mJy)	$S_{610MHz,\,k-corr}$ (mJy)	I_{610MHz} (μJy/arcsec^2)	P_{610MHz} (10^{24} W Hz^{-1})
A	10.0 ± 2.0	12.0 ± 2.0	6.0 ± 1.0	9.0 ± 2.0
B	19.0 ± 3.0	–	11.0 ± 2.0	–
C	11.0 ± 2.0	–	6.3 ± 0.9	–
D	5.2 ± 0.8	4.7 ± 0.7	3.0 ± 0.4	3.4 ± 0.5

k-corrected flux and P_{601MHz} are calculated assuming a spectral index of 0.7 for region D and 1.4 for region A

3.3.2 *MACS J2243.3-0935 at <1 GHZ*

Figures 3.4 and 3.5 show diffuse radio emission detected in MACS J2243.3-0935 by the GMRT at 610 MHz. In the GMRT images, the diffuse emission detected by KAT-7 is resolved into four distinct regions labelled A to D in Fig. 3.5. Table 3.3 lists the flux densities for each of these regions. Flux densities were measured from the naturally weighted uvtapered image from within the 3σ contour level. Errors in flux measurements were calculated using the formula:

$$\sigma_{S_{610}} = \sqrt{(\sigma_{\rm cal} S_{610})^2 + \left(\sigma_{\rm rms}\sqrt{N_{\rm beam}}\right)^2}, \tag{3.1}$$

where $\sigma_{\rm cal}$ is the uncertainty in the calibration of the flux-scale and $N_{\rm beam}$ is the number of independent beams in the source. $\sigma_{\rm cal}$ is taken to be 10% for the GMRT (Chandra et al. 2004). Figure 3.6 shows the high resolution GMRT image of MACS J2243.3-0935 used to subtract the point sources with contours of the GMRT tapered image and KAT-7 image overlaid. The flux measured from the high resolution GMRT image within the same region of the KAT-7 cluster emission at a 3σ level is approximately 63 mJy. Extrapolating this flux to 1826 MHz assuming a spectral index of $\alpha = 0.7$, where $S_\nu \propto \nu^{-\alpha}$, gives a value of approximately 29 mJy. Subtracting this from the KAT-7 flux calculated in Sect. 3.3.1 leaves a residual flux of 11 mJy. The total flux measured from the point source subtracted, tapered GMRT image within the same region of the KAT-7 cluster emission is approximately 40 mJy. This suggests that the average spectral index of the diffuse emission in MACS J2243.3-0935 is 1.1.

3.3.2.1 Field Sources

In order to further examine the flux scale, the GMRT data were imaged with full uvrange, a Briggs weighting robust parameter of 0 and tapered close to the NVSS

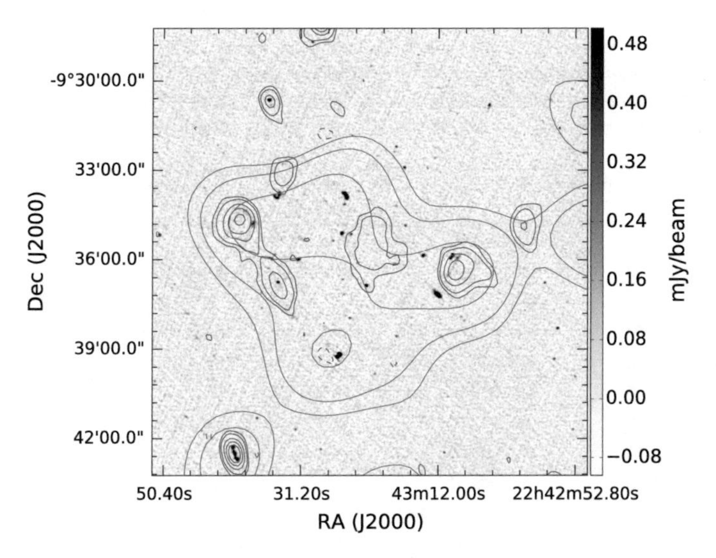

Fig. 3.6 Greyscale image shows the robust 0 high resolution GMRT data used for the point source subtraction. KAT-7 contours are overlaid in blue while contours for the tapered, point source subtracted GMRT image are overlaid in red. Contours are at 3, 5, 10, 15, 20 × σ_{rms} where $\sigma_{\text{rms}} = 500$ μJy/beam for KAT-7 and −3 3, 5, 10, 15, 20 × σ_{rms} $\sigma_{\text{rms}} = 200$ μJy/beam for the GMRT. The resolution of the KAT-7 image is 160.10 × 144.99 arcsec. The resolution of the high resolution GMRT image is 7.44 × 6.06 arcsec while the resolution of the tapered GMRT image is 44.89 × 33.70 arcsec

resolution. The software PYBDSF[3] was used to detect sources in both the NVSS and the GMRT maps. PYBDSF works by detecting all pixels in the map above a set peak threshold. It will then form islands of contiguous emission down to a set island threshold around the identified peak pixels. Gaussians are then fit to the islands and the Gaussians are grouped into individual sources. The flux densities of the sources are calculated by summing the flux densities of the Gaussians and the error in the flux density is calculated by summing the uncertainties in the Gaussians in quadrature. For both the GMRT and the NVSS images, the peak threshold was set to 7σ and the island threshold was set to 5σ. Table 3.4 lists the detected sources and their flux densities in the GMRT and the NVSS maps. Table 3.4 also shows the calculated spectral index of each source that is detected in both maps. Figure 3.7 shows the spectral indices for sources in the field versus their flux densities. The average spectral index is 0.2 ± 0.6. An average spectral index of $\alpha = 0.7$ is expected for most optically thin extragalactic radio sources due to the energy distribution of cosmic rays produced in shocks. However at low frequencies, the spectral index distribution of faint sources is expected to have a flat tail due to the flattening of blazar spectra at frequencies below 1 GHz. (Massaro et al. 2014).

[3]PYBDSF documentation: http://www.astron.nl/citt/pybdsm/.

Table 3.4 Sources within GMRT primary beam half power point. Column 1-4 list the RA and Dec values of each source as well as error in the positions. Column 5 lists the integrated flux density at 610 MHz, column 6 is the integrated flux density measured from NVSS, column 7 is the offset between the source position measured in GMRT and NVSS. Column 8 is the spectral index of the source

R.A. (h:m:s)	R.A. err. (s)	Dec. (d:m:s)	Dec. err. (arcsec)	$S_{610,GMRT}$ (mJy)	S_{NVSS} (mJy)	Offset (arcsec)	α
22:44:31.20	0.10	−09:41:07.80	2.80	2.18 ± 0.32	–	–	–
22:44:26.90	0.00	−09:26:36.40	0.30	33.91 ± 0.73	48.16 ± 1.22	3.1	−0.42 ± 0.09
22:44:29.60	0.10	−09:45:39.50	1.30	6.22 ± 0.44	6.37 ± 0.75	1.3	−0.03 ± 0.38
22:44:12.80	0.00	−09:34:22.60	0.50	13.09 ± 0.48	18.34 ± 0.99	1.4	−0.41 ± 0.18
22:44:12.60	0.10	−09:46:29.40	2.30	2.50 ± 0.34	4.63 ± 0.73	5.6	−0.74 ± 0.58
22:44:02.40	0.00	−09:51:18.40	0.10	165.66 ± 0.67	143.02 ± 1.32	19.2	0.18 ± 0.03
22:44:03.20	0.10	−09:28:32.70	0.90	9.96 ± 0.51	7.56 ± 0.82	6.0	0.33 ± 0.33
22:43:56.60	0.10	−09:25:26.00	1.60	3.79 ± 0.37	–	–	–
22:43:49.40	0.00	−09:54:06.20	0.20	103.19 ± 0.61	87.99 ± 1.16	1.5	0.19 ± 0.04
22:43:40.30	0.10	−09:42:31.70	1.40	6.06 ± 0.44	–	–	–
22:43:39.10	0.10	−09:34:42.30	1.50	10.41 ± 0.49	–	–	–
22:43:38.40	0.20	−09:44:28.40	3.30	2.85 ± 0.37	–	–	–
22:43:35.60	0.10	−09:30:42.60	2.40	2.92 ± 0.36	–	–	–
22:43:34.40	0.10	−09:33:41.30	2.90	3.69 ± 0.35	–	–	–
22:43:28.60	0.10	−09:28:00.50	1.30	7.62 ± 0.44	–	–	–
22:43:27.80	0.10	−09:47:10.00	2.50	1.91 ± 0.29	–	–	–
22:43:25.80	0.00	−09:39:11.40	0.50	16.16 ± 0.48	6.91 ± 0.90	3.2	1.02 ± 0.37
22:43:24.90	0.00	−09:33:49.50	0.90	20.67 ± 1.03	6.45 ± 0.77	2.3	1.40 ± 0.36
22:43:23.70	0.10	−09:50:52.60	2.80	3.10 ± 0.36	–	–	–
22:43:18.80	0.00	−09:18:54.40	0.30	22.85 ± 0.50	18.54 ± 0.99	1.5	0.25 ± 0.16
22:43:18.60	0.10	−09:44:52.20	1.00	8.76 ± 0.47	4.40 ± 0.63	2.2	0.83 ± 0.42
22:43:09.20	0.10	−09:36:09.60	1.20	21.34 ± 0.85	–	–	–
22:43:00.00	00.20	−09:34:56.20	3.90	2.89 ± 0.36	–	–	–
22:42:53.50	0.10	−09:22:49.10	2.30	3.33 ± 0.35	–	–	–
22:42:16.30	0.10	−09:31:53.10	2.40	2.49 ± 0.35	–	–	–

3.3.2.2 Optical Counterparts in Regions A-D

Figure 3.8 shows the robust 0 images of regions A, B, C and D as well as the optical SDSS images of each region. The locations of discrete radio sources are marked. Galaxies within the redshift slice $z \pm 0.04\,(1+z)$ are deemed to be associated with the cluster (Wen et al. 2009). Table 3.5 lists the radio sources found in each region and their optical counterparts. An optical source was deemed to be the radio source's counterpart if the optical source was within one FWHM of the GMRT beam from the radio source. If more than one optical source lies with a FWHM of the radio source, then the source closest to the centroid of the radio source is deemed to be the optical counterpart.

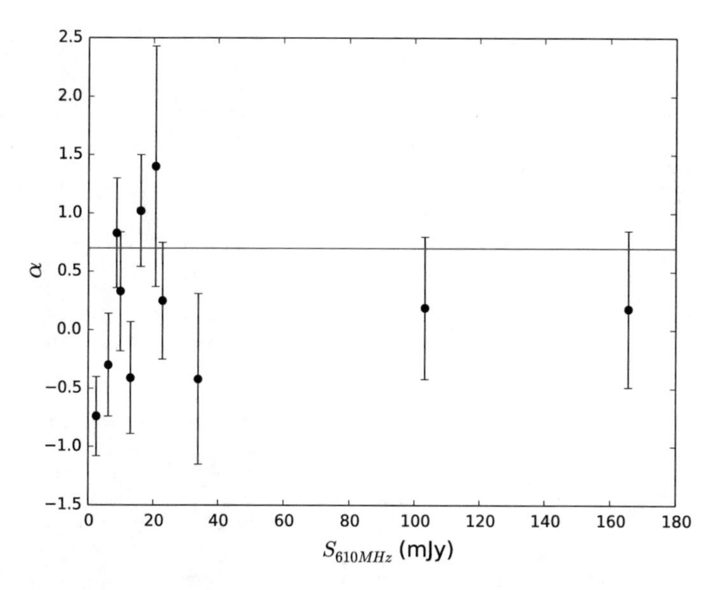

Fig. 3.7 Spectral index of sources versus their flux density within the primary beam half power point. The blue line marks a spectral index of 0.7

3.3.2.3 Region A

Region A is detected in the tapered image shown in Fig. 3.5 at a 5σ level. It is not detected at a significant level in either the untapered image, shown in Fig. 3.4, or the high resolution image used to subtract the point sources. The emission at the 3σ level fills roughly the same region as the X-ray emission, shown in Fig. 3.5. The radio emission appears to be extended along an axis almost perpendicular to the extension of the X-ray emission. In Fig. 3.8a there are no compact radio sources coincident with or near the peak of region A and so the diffuse emission is unlikely to be associated with a single discrete source. At a redshift 0.447, region A has a largest linear scale (LLS) of approximately 0.92 Mpc.

3.3.2.4 Region B

To the east of the cluster, complex diffuse emission can be seen in both the tapered and untapered image. The emission has a LLS of approximately 1.7 Mpc. In Fig. 3.8c there are two peaks in the emission. The southern peak is coincident with a discrete radio source, B-4, at J22:43:34.4 -09:35:58.6. There is no SDSS, X-ray or infra-red counterpart for this source. It is possible that the optical source has a high redshift that puts it outside the range of SDSS. This would place the source behind the cluster.

The northern peak is centred near source B-1 at J22:43:37.8 -9:34:46, which is located within the cluster. In Fig. 3.9 the northern peak is resolved into an arc of

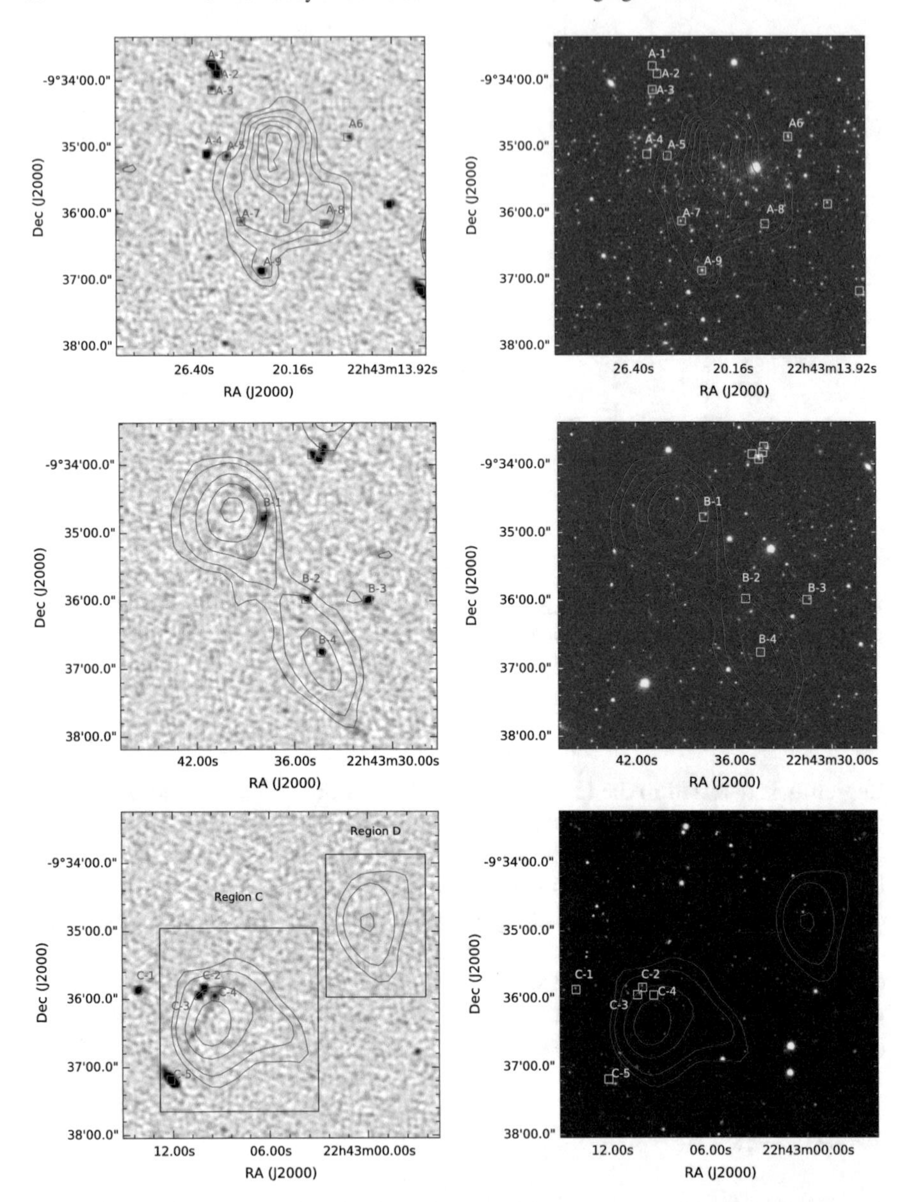

Fig. 3.8 Greyscale images show the robust 0 high resolution GMRT data in the left column and the rgb image of SDSS D12 i, r and g filters in the right column. The high resolution GMRT data has a rms noise of 40 μJy/beam and a resolution of 4.84 × 4.15 arcsec. The uvtapered GMRT images are overlaid in each image. The resolution of the uvtapered GMRT image is 44.89 × 33.70 arcsec. Radio point sources have been subtracted from the uvtapered image using models extracted from the high resolution GMRT image shown in the greyscale. The locations of discrete radio sources detected by PYBDSF are marked by blue boxes in the left column and by white boxes in the right column. The first row shows images for region A with contours are at −3, 3, 4, 5, 6, 7, 8, 9, 10, 15, 20 $\times\sigma_{\rm rms}$ where $\sigma_{\rm rms} = 200$ μJy/beam. The middle row shows images for region B and the last row shows images for regions C and D with contours at −3, 3, 5, 10, 15, 20 $\times\sigma_{\rm rms}$ where $\sigma_{\rm rms} = 200$ μJy/beam

Table 3.5 Sources within different regions. Column 1 is the region name and column 2 is the source name. Column 3 and 4 are the position of the sources in RA and Dec. Column 5 is the integrated flux density at 610 MHz, column 6 is the name of the SDSS optical counterpart, column 7 is the offset between the position of the GMRT source and the SDSS counterpart. Column 8 is the sources position relative to the cluster

Region	Source	RA (h:m:s)	Dec (d:m:s)	S_{610} (mJy)	SDSS source	Offset (arcsec)	Position
Region A	A-1	22:43:25.1	−09:33:46.8	3.62 ± 0.07	SDSS J224324.84-093350.9	3.05	Cluster member
	A-2	22:43:24.8	−09:33:54.0	5.18 ± 0.07	SDSS J224324.84-093350.9	3.05	Cluster member
	A-3	22:43:25.1	−09:34:08.5	0.61 ± 0.07	SDSS J224325.14-093408.7	0.25	Cluster member
	A-4	22:43:25.5	−09:35:06.8	1.94 ± 0.07	SDSS J224325.30-093503.1	4.68	Cluster member
	A-5	22:43:24.2	−09:35:08.3	0.52 ± 0.08	–	–	–
	A-6	22:43:16.6	−09:34:51.3	0.48 ± 0.08	SDSS J224316.63-093451.4	0.17	Foreground source
	A-7	22:43:23.3	−09:36:07.2	0.37 ± 0.08	SDSS J224323.40-093607.5	1.17	Cluster member
	A-8	22:43:18.1	−09:36:09.8	0.84 ± 0.06	–	–	–
	A-9	22:43:22.1	−09:36:52.2	2.02 ± 0.08	SDSS J224322.08-093652.4	0.18	Foreground source
Region B	B-1	22:43:37.8	-09:34:46.6	2.73 ± 0.05	SDSS J224337.71-093444.5	2.76	Cluster member
	B-2	22:43:35.3	−09:35:58.7	1.47 ± 0.07	–	–	–
	B-3	22:43:31.5	−09:35:59.7	1.46 ± 0.07	SDSS J224331.40-093558.9	1.47	Cluster member
	B-4	22:43:34.4	−09:36:45.6	1.24 ± 0.08	–	–	–
Region C	C-1	22:43:14.1	-09:35:52.1	1.71 ± 0.07	SDSS J224314.19-093551.1	1.12	Cluster member
	C-2	22:43:10.1	−09:35:50.3	1.56 ± 0.07	SDSS J224310.12-093548.7	1.74	Foreground source
	C-3	22:43:10.4	−09:35:56.8	1.50 ± 0.07	SDSS J224310.28-093555.4	2.09	Cluster member
	C-4	22:43:09.4	−09:35:57.1	1.31 ± 0.06	–	–	–
	C-5	22:43:12.1	−09:37:10.8	6.83 ± 0.08	SDSS J224311.92-093714.8	5.15	Cluster member

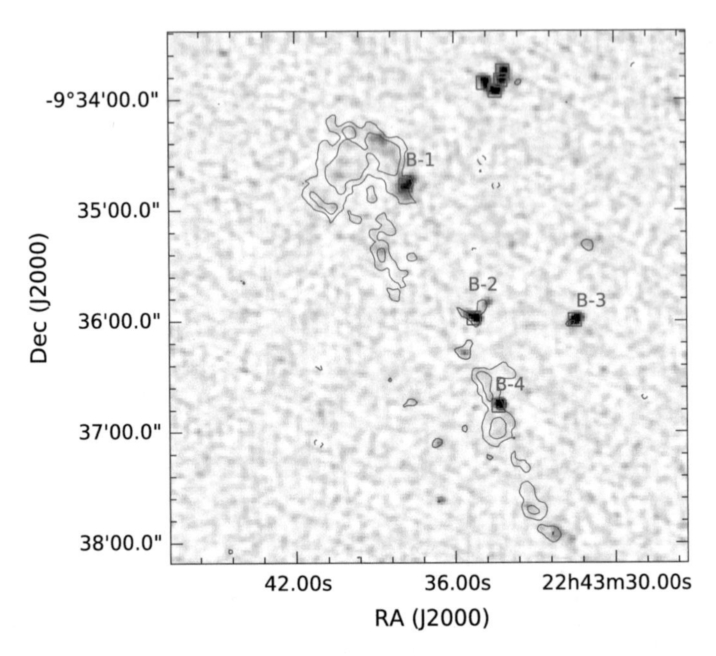

Fig. 3.9 Greyscale images show the robust 0 high resolution GMRT data for region B. Red contours show naturally weighted image. Contours are at −3, 3, 5, 10, 15, 20 × σ_{rms} where $\sigma_{\text{rms}} = 45$ μJy/beam. The resolution of the GMRT image is 7.44 × 6.06 arcsec. Radio point sources have been subtracted from the naturally weighted image using models extracted from the high resolution GMRT image shown in the greyscale. The blue squares mark discrete radio sources

emission with one end of the arc coincident with B-1. In the untapered image there is no emission detected connecting the northern and southern areas of region B.

3.3.2.5 Regions C and D

To the west of the cluster, there is a second region of complex diffuse emission. Again this can be seen in both the tapered and untapered maps. There appear to be two separate sources. The first, region C, is brighter and appears to coincident with at least two discrete sources, C-2 and C-3. A narrow linear structure is evident in the highest resolution greyscale image as well as the untapered image of region C in Fig. 3.10. The linear structure extends from the north-west to south-east. Region C has a LLS of approximately 0.76 Mpc.

The second region, region D, is on the edge of the cluster's virial radius. There are no discrete radio or optical sources evident in the region. It has a LLS of approximately 0.68 Mpc. The eastern side of region D is curved while the western side of region D is somewhat flatter.

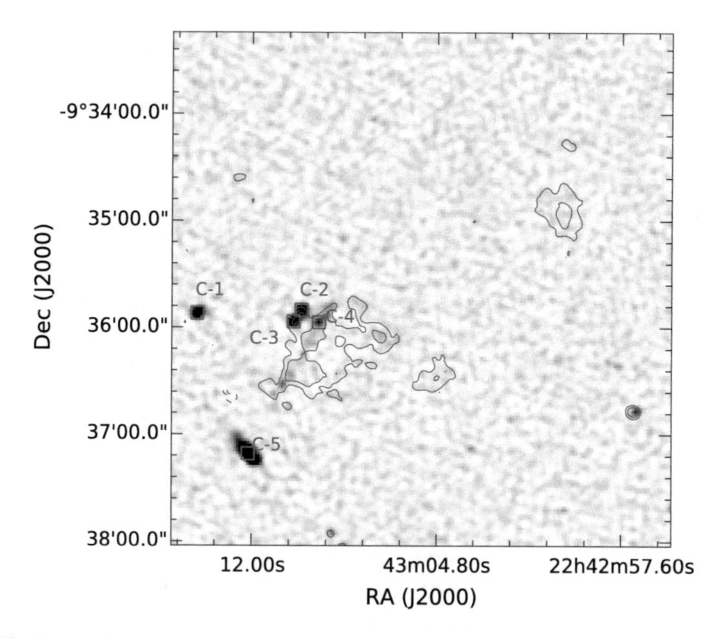

Fig. 3.10 Greyscale images show the robust 0 high resolution GMRT data for regions C and D. Red contours show naturally weighted image. Contours are at −3, 3, 5, 10, 15, 20 × $\sigma_{\rm rms}$ where $\sigma_{\rm rms} = 45\,\mu$Jy/beam. The resolution of the GMRT image is 7.44 × 6.06 arcsec. Radio point sources have been subtracted from the naturally weighted image using models extracted from the high resolution GMRT image shown in the greyscale. The blue squares mark discrete radio sources

3.4 Discussion

3.4.1 A Giant Radio Halo in MACS J2243.3-0935

Figure 3.5 shows the X-ray emission of the cluster with the radio contours of region A overlaid. The morphology, size and position of region A are consistent with that of a giant halo.

3.4.1.1 Spectral Index

As discussed in Sect. 3.3.2.3, Region A is clearly detected in the GMRT 610 MHz image, however it is not detected in the NVSS image and in the KAT-7 image all cluster emission is unresolved. Without high resolution data at 1.822 GHz, it is not possible to disentangle emission in region A from emission in region B, C or D in the KAT-7 image. The NVSS image does not have the resolution or the sensitivity to subtract the discrete sources from the KAT-7 image. As such I am only able to put a lower limit on the spectral index of the radio halo using the NVSS image. The rms noise of NVSS is 0.45 mJy/beam. Thus a 3σ upper limit flux density for the radio halo at 1400 MHz is 1.35 mJy/beam. Assuming the halo has the same spatial extent at 1400 MHz this gives an integrated upper limit on integrated flux of 8.2 mJy. Taken with the 610 MHz flux density of 10.0 ± 2.0 mJy this gives a lower limit on the spectral index of $\alpha \geq 0.28$. Radio Halos are expected to have much steeper spectral indices than 0.28, however NVSS does not have the surface brightness sensitivity to more tightly constrain the spectral index of region A.

Feretti et al. (2012) suggest a link between the average temperature of a cluster and the spectral index of radio halos . They find that radio halos in clusters with an average temperature between 8 and 10 keV have an average spectral index of $\alpha = 1.4 \pm 0.4$. MACS J2243.3-0935 has an temperature of 8.24 ± 0.92 K and so an estimate spectral index of $\alpha = 1.4$ will be used in this paper to estimate the properties of the halo in MACS J2243.3.

3.4.1.2 Scaling Relations

Using the spectral index stated above, the k-corrected radio power of the halo at 610 MHz is $P_{610\text{MHz}} = 9.0 \pm 2.0 \times 10^{24}$ W Hz^{-1}. Figure 3.11 shows the halo's position on the $P_{610\,\text{MHz}} - L_{\text{x}}$ and $P_{610\,\text{MHz}} - M_{500}$ diagrams. Figure 3.11 is a reproduction of Figure 2 in Yuan et al. (2015) with the data point for MACS J2243.3-0935 included. Region A in MACS J2243.3-0935 is in good agreement with the power expected from the correlations show in Fig. 3.11, providing further evidence that region A is a radio halo.

3.4.1.3 Equipartition B-Fields

Beck and Krause (2005) provide a revised formula for the classical equipartition magnetic field,

$$B = \left\{ \frac{4\pi\,(2\alpha+1)\,(K_0+1)\,I_\nu E_p^{1-2\alpha}\left(\frac{\nu}{2c_1}\right)^{\alpha}}{(2\alpha-1)\,c_2 l c_4} \right\}^{\frac{1}{\alpha+3}}, \tag{3.2}$$

where α is the spectral index, K_0 is the ratio of proton energy densities to electron energy densities, I_ν is the synchrotron intensity, E_{p} is the proton rest energy, ν is the observing frequency and l is the extent of the source along the line of sight. c_1 and c_3 are constants while c_2 is a function of the spectral index and c_4 is a function of the inclination of the source. See Appendix A in Beck and Krause (2005) for definition of these variables.

There is much discussion in the literature on the precise value of K_0. Different CR injection mechanisms predict different values for K_0 for the ICM. For example, turbulent acceleration predicts $K_0 = 100$, production of secondary CRe predicts K_0 in the range of 100 to 300 and first order Fermi shock acceleration predicts values of

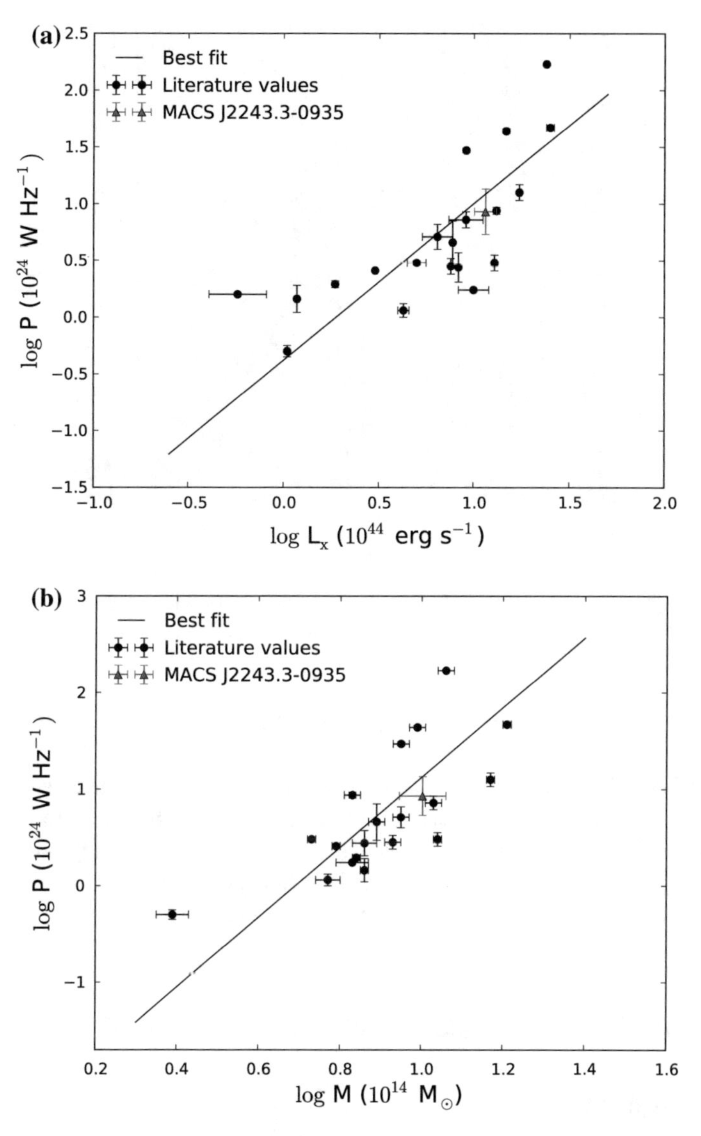

Fig. 3.11 These plots marks the position of the halo in MACS J2243.3-0925 in red on the 610 MHz scaling relations examined in Yuan et al. (2015) **a** shows the $P_{610\,\mathrm{MHz}} - L_x$ correlation and **b** shows the $P_{610\,\mathrm{MHz}} - M$ relation. Black data points are taken from Yuan et al. (2015)

K_0 in the range of 40 to 100 (Beck and Krause 2005). However energy losses such as synchrotron and inverse Compton could inflate the value of K_0 to values much greater than 100. Vazza and Brüggen (2014) compare some of the current models for CR injection to radio and Fermi data on clusters. They find that values of $K_0 \geq 100$

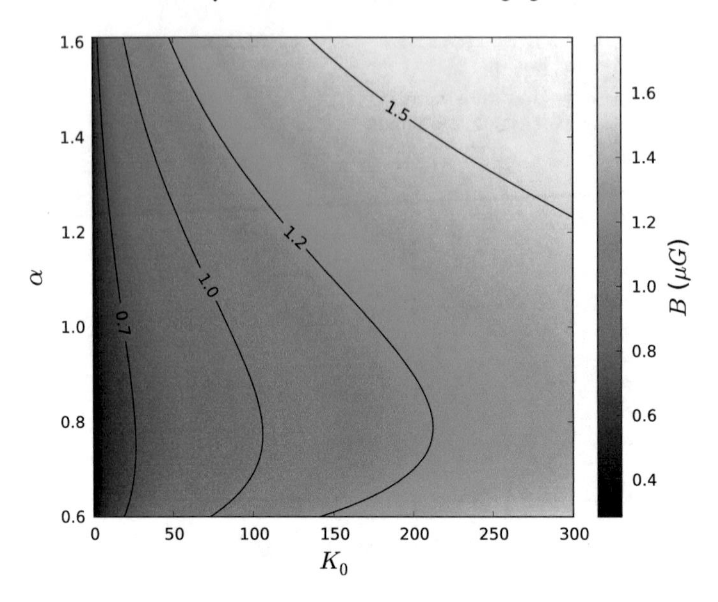

Fig. 3.12 Equipartition magnetic field strength for region A for a range of values of α and K_0. Black contours mark regions of constant magnetic field strength

require gamma-ray emission above the derived Fermi upper limits, suggesting that $K_0 \leq 100$.

Using Eq. 3.2, I calculate the equipartition magnetic field of the cluster in region A and region B for different values of α and K_0. Figure 3.12 shows the results for region A while Fig. 3.13 shows the results for region B. For both regions, magnetic field strengths vary from less than 0.5 μG for flat spectral indices and small values of K_0 to 1.5 μG for steep spectral indices and high values of K_0.

3.4.2 *Possible Radio Relics in MACS J2243.3-0935*

3.4.2.1 Region B

There are four possible explanations for the diffuse emission in Region B. The first is that it is merely a superposition of emission from sources at different redshifts. The second is that the emission is associated with the interaction of sources at the same redshift. The third is that the emission is from a giant radio galaxy (GRG). And finally the emission could be a radio relic.

The LLS of region B is consistent with both a GRG or a radio relic . The position of region B at the periphery of a cluster is expected for a radio relic while GRG are more commonly found in less dense regions such as galaxy groups (Malarecki et al. 2015). The double peaked morphology of region B is unusual for a radio relic.

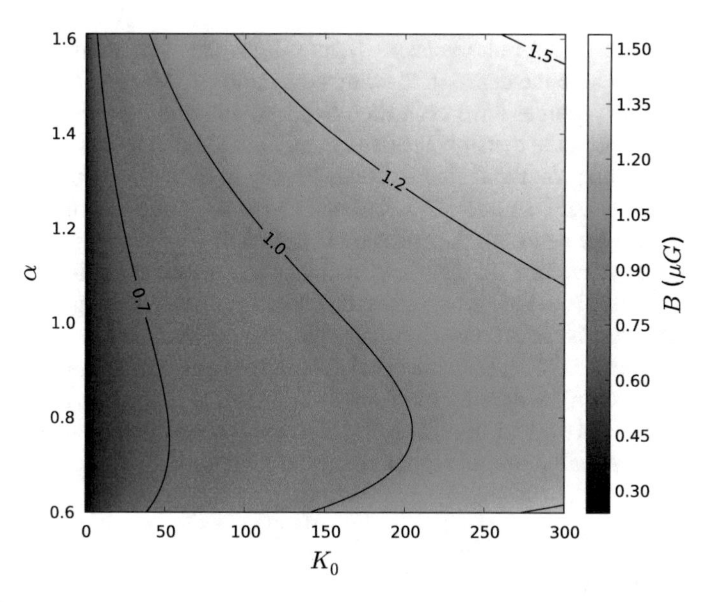

Fig. 3.13 Equipartition magnetic field strength for region D for a range of values of α and K_0. Black contours mark regions of constant magnetic field strength

A possible explanation for these localised regions of increased emission, in the context of a radio relic, is that these peaks coincide with areas of fossil AGN activity. The fossil CRe in these areas would be accelerated to higher energies and lead to a localised increase in emission.

The northern peak in Fig. 3.9 could also be interpreted as a bow shock in the lobe of a GRG . However it seems more likely that this is instead emission associated with B-1. B-1 could be a head-tail galaxy with the head located at B-1 and the tail curving north then north-east. Alternatively B-1 could be a wide angle tailed source with the emission south of B-1 forming the second jet/lobe.

While the linear size of region B is consistent with a radio relic or GRG , comparison between Figs. 3.4 and 3.5 suggest that region B is not a region of continuous emission, but instead a series of radio galaxies that are unresolved in the tapered image, thus ruling out a radio relic and GRG . This would place B-1 and B-4 as the sources of the northern and southern peaks respectively.

B-4 and B-2 do not have optical counterparts in SDSS. They are likely background galaxies outside the redshift range of SDSS, suggesting that region B is a superposition of sources at different redshifts.

3.4.2.2 Region C

Figure 3.10 shows the high resolution GMRT data for region C with the naturally weighted contours overlaid. The morphology of region C seen in Fig. 3.10 is consistent with bent tail radio galaxy . Bent tail galaxies are commonly found in galaxy clusters where the dense ICM warps the radio jet (Mao et al. 2009; Pratley et al. 2013). Such sources have been used to probe different physical properties of the ICM , including the density (Freeland et al. 2008) and magnetic field strength (Clarke et al. 2001; Vogt and Enßlin 2003; Pratley et al. 2013). However due the cluster of discrete radio sources, C-2, C-3 and C-4, a superposition of sources can't be ruled as an explanation for the emission in region C.

The position of region C in the cluster and its LLS are also consistent with that of a radio relic. There is evidence to suggest that some radio relics are generated when shocks re accelerated fossil plasma from radio galaxies (Enßlin and Gopal-Krishna 2001; Bonafede et al. 2014; Shimwell et al. 2015). In such a scenario, region B could be a produced when fossil emission from C-2, C-3 or C-4 was reaccelerated by a merger shock . However without knowing how the spectral index varies across the source it is not possible to differentiate between a radio relic and a bent tail radio galaxy.

3.4.2.3 Region D

NVSS and FIRST do not detect any radio sources in Region D. The rms noise of NVSS is 0.45 mJy/beam. Thus a 3σ upper limit flux density for region D at 1400 MHz is 1.35 mJy/beam. Assuming region D has the same spatial extent at 1400 MHz this gives an upper limit on the integrated flux density of 2.0 mJy. Taken with the 610 MHz flux density of 5.2mJy this gives a lower limit on the spectral index of $\alpha \geq 0.7$.

The greyscale image in Fig. 3.10 shows data from baselines longer than 4 kλ imaged with a Briggs robust parameter of 0 for both region C and region D with the contours for the naturally weighted, point source subtracted image overlaid. No compact radio sources can be seen in this image coincident with or near the peak of region D and so the diffuse emission is unlikely to be associated with a single discrete source.

With a LLS of 0.68 Mpc, Region D is consistent with that of a radio relic . The lack of radio point sources in region D suggests this emission is not associated with a discrete source. Region D is about twice the length on the north-south axis as on the east-west axis, which is consistent with the morphology of elongated radio relics . Radio relics are likely formed by shocks produced by either major/minor cluster mergers or through the infall of the warm-hot intergalactic medium (WHIM) onto the cluster. Shocks produced by cluster mergers are expected to have a Mach number less than 5 (Skillman et al. 2008). Hong et al. (2014) study the properties of shocks at cluster outskirts and suggest that around half of radio relics with Mach numbers greater than 3, as well as relatively flat radio spectra, are infall shocks . To date only a few relics have been described as infall relics in the literature. For example, Brown and Rudnick (2011) suggest that the radio relic 1253 + 275 in the Coma cluster is caused by the infalling group NGC 4839 while Pfrommer and Jones (2011) model the structure of the head tail radio galaxy NGC 1265 by assuming that the galaxy

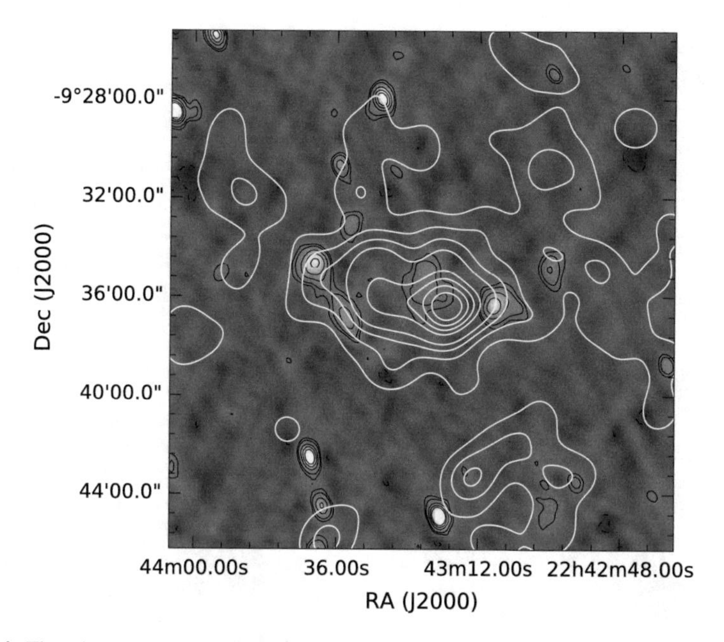

Fig. 3.14 These images show MACS J2243.3-0935 at different wavelengths with red contours overlaid showing the tapered image. Contours are at 3, 5, 10, 15, 20 $\times$ $\sigma_{\rm rms}$ where $\sigma_{\rm rms} = 200\,\mu$Jy/beam. **a** IRIS 25 μm **b** IRIS 60 μm **c** IRIS 100 μm **d** SHASS Hα **e** WISE 12 μm **f** WISE 22 μm

passed through an accretion shock onto the Perseus cluster. Pfrommer and Jones (2011) calculate the Mach number of the inferred accretion shock in the Perseus cluster to be approximately $\mathcal{M} = 4.2$.

MACS J2243.3-0935 is the central cluster in the supercluster SCL2243-0935. Figure 3.15 shows the number density of SDSS galaxies in the region of MACS J2243-0935. The galaxy density at each point was calculated by counting the number of galaxies that fell in each pixel. The resultant map was then smoothed with a Gaussian of width 42 arcsec. Galaxies were chosen to be in the photometric redshift bin $0.39 < z < 0.5$ and within 4 Mpc of the cluster centre. This redshift bin was chosen to account for uncertainties in the photometric redshift error and a large spatial region was chosen to include the filamentary structure in the supercluster. Region D is located at the virial radius where one of the supercluster filaments, AH in Schirmer et al. (2011), meets MACS J2243.3-0935. The location of region D as well as the LLS are suggestive of an infall relic , however multi-frequency analysis would be required to properly determine the nature of the region.

In order to rule out Galactic foreground emission as an explanation for region D, Fig. 3.14 shows region D at multiple wavelengths. There is no significant emission in IRIS, SHASSA Hα, WISE or *Planck*, and I conclude that region D is unlikely to be associated with Galactic foreground emission.

Given the coincidence of region D with a filament, the radio emission might alternatively be associated with the WHIM rather than a radio relic. There is no

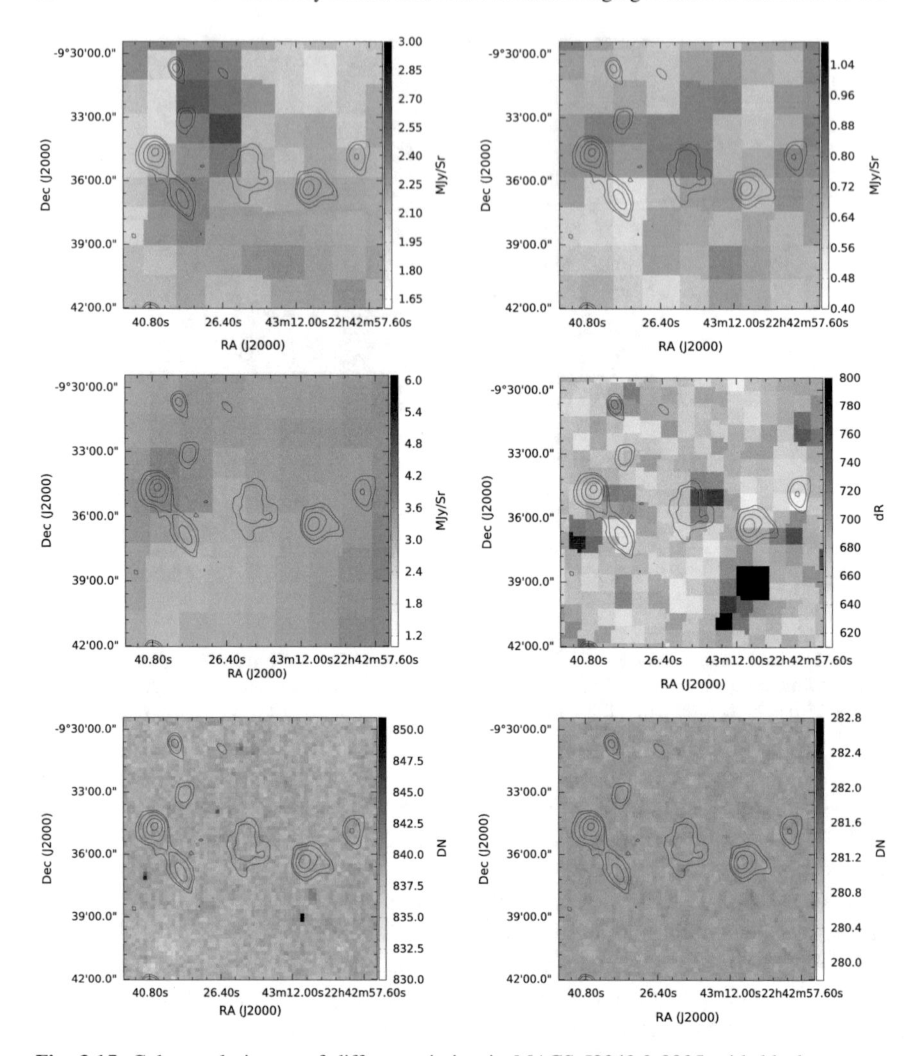

Fig. 3.15 Colourscale image of diffuse emission in MACS J2243.3-0935 with black contours overlaid showing the tapered image. Contours are at -3, 3, 5, 10, 15, 20 $\times$ $\sigma_{\rm rms}$ where $\sigma_{\rm rms}$ = 200 μJy/beam. Galaxy density contours overlaid in white. These contours are range from 20% to 90% of the peak value in steps of 10

agreement on precise predictions for magnetic fields in filaments but estimates range from 10^{-4} to 0.1 μG (Dolag et al. 1999; Brüggen et al. 2005; Ryu et al. 2008). In Sect. 3.4.1.3 the equipartition magnetic field estimates for region D for a spectral index of 0.7 are greater than the estimates for magnetic fields in the WHIM. Araya-Melo et al. (2012) model cosmic rays in large scale structure and predict a flux density of 0.12 μJy/beam at a frequency of 150 MHz for a 10 arcsec^2 beam. This is much lower than the flux density measured in region D or in other possible radio detections of the WHIM (Bagchi et al. 2002; Farnsworth et al. 2013). The high flux density and magnetic field estimates for region D suggest that it is unlikely to be a radio detection of the WHIM.

3.5 Conclusion

I have discovered a radio halo in the merging cluster MACS J2243.3-0935 using GMRT observations at 610 MHz and KAT-7 observations at 1822 MHz. The radio halo has an integrated flux density of $S_{610\mathrm{MHz}} = 10.0 \pm 2.0$ mJy, an estimated radio power at 1.4 GHz of $P_{1.4\mathrm{GHz}} = 3.2 \pm 0.6 \times 10^{24}$ WHz^{-1} and a LLS of approximately 0.92 Mpc. I calculated the equipartition magnetic field in the region of the halo for a range of α and K_0 values and find that the equipartition magnetic field is of order 1 μG. Assuming a spectral index of $\alpha = 1.4$, the halo in MACS J2243.3-0935 lies on the empirical scaling relations observed for radio halos.

I also detected a potential radio relic candidate to the west of the cluster. The candidate relic has a integrated flux density of 5.2 ± 0.8 mJy, an estimated radio power at 1.4 GHz of $(1.6 \pm 0.3) \times 10^{24}$ W Hz^{-1} and a LLS of 0.68 Mpc. The presence of a radio relic in MACS J2243.3-0935 would make this one of only a handful of clusters that host both a halo and a relic. Due to the position of the relic candidate on the outskirts of the cluster, where a filament meets the cluster, I conclude that the candidate is consistent with an infall relic. I rule out the possibility of the emission being associated with the WHIM in a filament as the measured flux density and estimated equipartition magnetic field strength are both much larger than expected values for the WHIM. I also exclude foreground galactic emission as an explanation as there is no significant emission in IRIS, SHASSA Hα, WISE or *Planck*.

References

Araya-Melo PA, Aragón-Calvo MA, Brüggen M, Hoeft M (2012) Radio emission in the cosmic web. MNRAS 423:2325–2341. https://doi.org/10.1111/j.1365-2966.2012.21042.x, arXiv:1204.1759

Bagchi J, Enßlin TA, Miniati F, Stalin CS, Singh M, Raychaudhury S, Humeshkar NB (2002) Evidence for shock acceleration and intergalactic magnetic fields in a large-scale filament of galaxies ZwCl 2341.1+0000. New A 7:249–277. https://doi.org/10.1016/S1384-1076(02)00137-9, arXiv:astro-ph/0204389

Becker RH, White RL, Helfand DJ (1995) The FIRST survey: faint images of the radio sky at twenty centimeters. APJ 450:559. https://doi.org/10.1086/176166

Beck R, Krause M (2005) Revised equipartition and minimum energy formula for magnetic field strength estimates from radio synchrotron observations. Astronomische Nachrichten 326:414–427. https://doi.org/10.1002/asna.200510366, arXiv:astro-ph/0507367

Bock DCJ, Large MI, Sadler EM (1999) SUMSS: a wide-field radio imaging survey of the southern sky. I. Sci Goals Surv Des Instrum. AJ 117:1578–1593, https://doi.org/10.1086/300786, arXiv:astro-ph/9812083

Bonafede A, Intema HT, Brüggen M, Girardi M, Nonino M, Kantharia N, van, Weeren RJ, Röttgering HJA, (2014) Evidence for Particle Re-acceleration in the radio relic in the galaxy cluster PLCKG287.0+32.9. APJ 785:1. https://doi.org/10.1088/0004-637X/785/1/1, arXiv:1402.1492

Brown S, Rudnick L (2011) Diffuse radio emission in/around the Coma cluster: beyond simple accretion. MNRAS 412:2–12. https://doi.org/10.1111/j.1365-2966.2010.17738.x, arXiv:1009.4258

Brüggen M, Ruszkowski M, Simionescu A, Hoeft M, Dalla Vecchia C (2005) Simulations of magnetic fields in filaments. APJl 631:L21–L24. https://doi.org/10.1086/497004, arXiv:astro-ph/0508231

Cantwell TM, Scaife AMM, Oozeer N, Wen ZL, Han JL (2016) A newly discovered radio halo in merging cluster MACS J2243.3-0935. MNRAS 458:1803–1814. https://doi.org/10.1093/mnras/stw419

Chandra P, Ray A, Bhatnagar S (2004) The late-time radio emission from SN 1993J at meter wavelengths. APJ 612:974–987. https://doi.org/10.1086/422675, arXiv:astro-ph/0405448

Clarke TE, Kronberg PP, Böhringer H (2001) A New radio-X-Ray probe of galaxy cluster magnetic fields. APJl 547:L111–L114. https://doi.org/10.1086/318896, arXiv:astro-ph/0011281

Condon JJ, Cotton WD, Greisen EW, Yin QF, Perley RA, Taylor GB, Broderick JJ (1998a) The NRAO VLA Sky survey. AJ 115:1693–1716. https://doi.org/10.1086/300337

De Gasperin F, Intema HT, Williams W, Brüggen M, Murgia M, Beck R, Bonafede A (2014) The diffuse radio emission around NGC 5580 and NGC 5588. Mon Not Royal Astron Soc 440:1542–1550. https://doi.org/10.1093/mnras/stu360, http://arxiv.org/abs/1402.5528, http://www.arxiv.org/pdf/1402.5528.pdf, arXiv:1402.5528

Dolag K, Bartelmann M, Lesch H (1999) SPH simulations of magnetic fields in galaxy clusters. A&A 348:351–363 astro-ph/0202272

Ebeling H, Edge AC, Mantz A, Barrett E, Henry JP, Ma CJ, van Speybroeck L (2010) The x-ray brightest clusters of galaxies from the massive cluster survey. Mon Not Royal Astron Soc 407(1):83–93. https://doi.org/10.1111/j.1365-2966.2010.16920.x, http://mnras.oxfordjournals.org/content/407/1/83.abstracthttp://mnras.oxfordjournals.org/content/407/1/83.full.pdf+html

Enßlin TA, Gopal-Krishna, (2001) Reviving fossil radio plasma in clusters of galaxies by adiabatic compression in environmental shock waves. A&A 366:26–34. https://doi.org/10.1051/0004-6361:20000198, arXiv:astro-ph/0011123

Farnsworth D, Rudnick L, Brown S, Brunetti G (2013) Discovery of Megaparsec-scale, low surface brightness nonthermal emission in merging galaxy clusters using the green bank telescope. APJ 779:189. https://doi.org/10.1088/0004-637X/779/2/189, arXiv:1311.3313

Feretti L, Giovannini G, Govoni F, Murgia M (2012) Clusters of galaxies: observational properties of the diffuse radio emission. Astron Astrophys Rev 20(1):54, https://doi.org/10.1007/s00159-012-0054-z, https://doi.org/10.1007/s00159-012-0054-z, arXiv:1205.1919v1

Freeland E, Cardoso RF, Wilcots E (2008) Bent-double radio sources as probes of intergalactic gas. APJ 685:858–862. https://doi.org/10.1086/591443, arXiv:0806.3971

Helmboldt JF, Kassim NE, Cohen AS, Lane WM, Lazio TJ (2008) Radio frequency spectra of 388 bright 74 MHz sources. APJs 174:313–336. https://doi.org/10.1086/521829, arXiv:0707.3418

Hong SE, Ryu D, Kang H, Cen R (2014) Shock waves and cosmic ray acceleration in the outskirts of galaxy clusters. APJ 785:133. https://doi.org/10.1088/0004-637X/785/2/133, arXiv:1403.1420

Malarecki JM, Jones DH, Saripalli L, Staveley-Smith L, Subrahmanyan R (2015) Giant radio galaxies—II. Tracers of large-scale structure. MNRAS 449:955–986. https://doi.org/10.1093/mnras/stv273, arXiv:1502.03954

Mann AW, Ebeling H (2012) X-ray optical classification of cluster mergers and the evolution of the cluster merger fraction. Mon Not Royal Astron Soc 420(3):2120–2138. https://doi.org/10.1111/j.1365-2966.2011.20170.x, http://mnras.oxfordjournals.org/content/420/3/2120.abstract, http://mnras.oxfordjournals.org/content/420/3/2120.full.pdf+html

Mantz A, Allen SW, Ebeling H, Rapetti D, Drlica-Wagner A (2010) The observed growth of massive galaxy clusters ii. x-ray scaling relations. Mon Not Royal Astron Soc 406(3):1773–1795. https://doi.org/10.1111/j.1365-2966.2010.16993.x, http://mnras.oxfordjournals.org/content/406/3/1773.abstract, http://mnras.oxfordjournals.org/content/406/3/1773.full.pdf+html

Mao MY, Johnston-Hollitt M, Stevens JB, Wotherspoon SJ (2009) Head-tail Galaxies: beacons of high-density regions in clusters. MNRAS 392:1070–1079. https://doi.org/10.1111/j.1365-2966.2008.14141.x, arXiv:0810.4739

Massaro F, Giroletti M, D'Abrusco R, Masetti N, Paggi A, Cowperthwaite PS, Tosti G, Funk S (2014) The low-frequency radio catalog of flat-spectrum sources. APJs 213:3. https://doi.org/10.1088/0067-0049/213/1/3, arXiv:1503.03483

McMullin JP, Waters B, Schiebel D, Young W, Golap K (2007) CASA architecture and applications. In: Shaw RA, Hill F, Bell DJ (eds) astronomical data analysis software and systems XVI, Astronomical Society of the Pacific Conference Series, vol 376, p 127

Offringa AR, van de Gronde JJ, Roerdink JBTM (2012) A morphological algorithm for improved radio-frequency interference detection. A&A 539

Pauliny-Toth IIK, Wade CM, Heeschen DS (1966) Positions and flux densities of radio sources. APJs 13:65. https://doi.org/10.1086/190137

Perley RA, Butler BJ (2013) An accurate flux density scale from 1 to 50 GHz. APJs 204:19. https://doi.org/10.1088/0067-0049/204/2/19, arXiv:1211.1300

Pfrommer C, Jones TW (2011) Radio galaxy NGC 1265 unveils the accretion shock onto the Perseus galaxy cluster. APJ 730:22. https://doi.org/10.1088/0004-637X/730/1/22, arXiv:1004.3540

Planck Collaboration, Ade PAR, Aghanim N, Armitage-Caplan C, Arnaud M, Ashdown M, Atrio-Barandela F, Aumont J, Aussel H, Baccigalupi C et al (2014) Planck 2013 results. XXIX. The Planck catalogue of Sunyaev-Zeldovich sources. A&A 571:A29, https://doi.org/10.1051/0004-6361/201321523, arXiv:1303.5089

Planck Collaboration, Ade PAR, Aghanim N, Arnaud M, Ashdown M, Aumont J, Baccigalupi C, Banday AJ, Barreiro RB, Barrena R et al (2015) Planck 2015 results. XXVII. The Second Planck Catalogue of Sunyaev-Zeldovich Sources. ArXiv e-prints arXiv:1502.01598

Pratley L, Johnston-Hollitt M, Dehghan S, Sun M (2013) Using head-tail galaxies to constrain the intracluster magnetic field: an in-depth study of PKS J0334–3900. MNRAS 432:243–257. https://doi.org/10.1093/mnras/stt448, arXiv:1303.2847

Rengelink RB, Tang Y, de Bruyn AG, Miley GK, Bremer MN, Roettgering HJA, Bremer MAR (1997) The westerbork northern sky survey (WENSS), I. A 570 square degree mini-survey around the north ecliptic pole. A&AS 124:259–280. https://doi.org/10.1051/aas:1997358

Ryu D, Kang H, Cho J, Das S (2008) Turbulence and magnetic fields in the large-scale structure of the Universe. Science 320:909, https://doi.org/10.1126/science.1154923, arXiv:0805.2466

Schirmer M, Hildebrandt H, Kuijken K, Erben T (2011) Mass, light and colour of the cosmic web in the supercluster SCL2243-0935 (z = 0.447). A&A 532:A57, https://doi.org/10.1051/0004-6361/201016348, arXiv:1102.4617

Shimwell TW, Markevitch M, Brown S, Feretti L, Gaensler BM, Johnston-Hollitt M, Lage C, Srinivasan R (2015) Another shock for the Bullet cluster, and the source of seed electrons for radio relics. MNRAS 449:1486–1494. https://doi.org/10.1093/mnras/stv334, arXiv:502.01064

Skillman SW, O'Shea BW, Hallman EJ, Burns JO, Norman ML (2008) Cosmological shocks in adaptive mesh refinement simulations and the acceleration of cosmic rays. APJ 689:1063–1077. https://doi.org/10.1086/592496, arXiv:0806.1522

Swarup G, Ananthakrishnan S, Kapahi VK, Rao AP, Subrahmanya CR, Kulkarni VK (1991) The giant Metre-Wave radio telescope. Curr Sci, 60, NO2/JAN25, 60:95

Vazza F, Brüggen M (2014) Do radio relics challenge diffusive shock acceleration? MNRAS 437:2291–2296. https://doi.org/10.1093/mnras/stt2042, arXiv:1310.5707

Vogt C, Enßlin TA (2003) Measuring the cluster magnetic field power spectra from Faraday rotation maps of Abell 400, Abell 2634 and Hydra A. A&A 412:373–385. https://doi.org/10.1051/0004-6361:20031434, arXiv:astro-ph/0309441

Wen ZL, Han JL (2013) Substructure and dynamical state of 2092 rich clusters of galaxies derived from photometric data. MNRAS 436:275–293. https://doi.org/10.1093/mnras/stt1581, arXiv:1307.0568

Wen ZL, Han JL, Liu FS (2009) Galaxy clusters identified from the SDSS DR6 and their properties. APJs 183:197–213. https://doi.org/10.1088/0067-0049/183/2/197, arXiv:0906.0803

Wright EL (2006) A cosmology calculator for the world wide web. PASP 118:1711–1715. https://doi.org/10.1086/510102, astro-ph/0609593

Yuan ZS, Han JL, Wen ZL (2015) The scaling relations and the fundamental plane for radio halos and relics of galaxy clusters. APJ 813(1):77, http://stacks.iop.org/0004-637X/813/i=1/a=77

Chapter 4
GMRT Observations of Three Highly Disturbed Clusters

4.1 Introduction

Diffuse radio emission in galaxy clusters, in the form of radio halos or relics, provides evidence that cosmic ray electrons (CRe) are present in the ICM, as are cluster-scale magnetic fields. Such sources provide a unique opportunity to study the non-thermal component of the ICM as well as an indirect method to probe the dynamics and evolution of galaxy clusters.

As discussed in Sect. 1.1.2, the reacceleration model is currently the preferred model for describing the formation of radio halos. However there are still many uncertainties in this model, in particular the acceleration efficiencies (Brunetti 2016). The amount of turbulent energy expected to be available to reaccelerate electrons is a function of cluster mass (Cassano 2010). As turbulent acceleration is an inefficient process a high energy cut off in the halo spectra is expected (Brunetti and Jones 2014). Thus halo s in low mass merging clusters would be expected to be only visible at lower frequencies (Cassano 2010; Cassano et al. 2012). Most detailed cluster studies have been carried out at $\nu = 1400$ MHz which samples only the most massive clusters and rarest mergers (Feretti et al. 2012). Observations at low frequencies ($\nu < 1400$ MHz) are necessary to properly sample the wider radio halo population.

In this chapter I present observations at 610 MHz of the dynamically disturbed clusters A07, A1235 and A2055. These clusters were classified as disturbed based on their optical relaxation parameter. Wen and Han (2013) define the relaxation parameter, Γ, of a galaxy cluster based on the amount of substructure evident in the cluster's optical properties (see Sect. 1.1.1). The relaxation parameter could prove a powerful tool for identifying possible host clusters of radio halos and was used to select MACS J2243.3-0935 for observation, where a new halo was presented in Sect. 2.6.4.

Sources were selected based on their X-ray luminosity and relaxation parameters. Sources were required to have a relaxation parameter of $\Gamma < -1$, to ensure highly disturbed clusters were selected, and an X-ray luminosity at $0.1 - 2.4$ keV

T. Cantwell, *Low Frequency Radio Observations of Galaxy Clusters and Groups*, Springer Theses, https://doi.org/10.1007/978-3-319-97976-2_4

Table 4.1 Properties of A07, A2055 and A1235

Cluster	A07	A2055	A1235	Reference
z	0.1073	0.101964	0.104200	Piffaretti et al. (2011)
$L_{x,500}$ (10^{44}erg s^{-1})	4.52	3.789308	1.696468	Böhringer et al. (2000)
Γ	−1.09	−1.55	−1.46	Wen and Han (2013)
r_{500} (Mpc)	1.0232	0.9979	0.8390	Piffaretti et al. (2011)

Table 4.2 GMRT observations summaries for A07, A1235 and A2055

	A07	A1235	A2055
Date	6 Mar 2016	4 Mar 2016	4 Mar 2016
Flux calibrator	3C48	3C286	3C286
Phase calibrator	0029 + 349	1051+213	1445 + 099
Time on target (hrs)	6.8	5.9	6.6

of $> 1 \times 10^{4}4$ erg s^{-1}. The Base de Donnes Amas de Galaxies X (BAX) catalogue (Sadat et al. 2004) was used to cross reference the X-ray luminosities for the clusters presented in Wen and Han (2013). Clusters were also required to have no previous discovered radio halo. The clusters selected for observation were A07, A1235 and A2055. Table 4.1 lists some of the properties of A07, A1235 and A2055. Like MACS J2243.3-0935, A07, A1235 and A2005 all have highly negative relaxation parameters. However the X-ray luminosity in these clusters is lower than in MACS J2243.3-0935 or indeed most clusters observed at 1400 MHz. This makes them ideal candidates for observation with low frequency telescopes.

4.2 Observations and Data Reduction

Observations were taken with the GMRT at 610 MHz on the 4th, 5th and 6th of March 2016. For each source a calibrator was observed for 10 min at the begining and end of the observation run. A phase calibrator was observed for five minutes every 10 min. Table 4.2 summarises the details of each observation.

Data were reduced using the SPAM pipeline (Intema 2014). SPAM is a direction dependent calibration pipeline implemented in AIPS. SPAM operates in three stages.

Table 4.3 Images properties for A07, A1235 and A2055

Cluster	High res. beam	High res. σ_{rms}	Low res. beam	Low res. σ_{rms}
A07	$5'' \times 5''$	35 μJy/beam	$56'' \times 55''$	0.55 mJy/beam
A2055	$5'' \times 5''$	50 μJy/beam	$59'' \times 55''$	0.5 mJy/beam
A1235	$5'' \times 5''$	30 μJy/beam	$59'' \times 56''$	0.25 mJy/beam

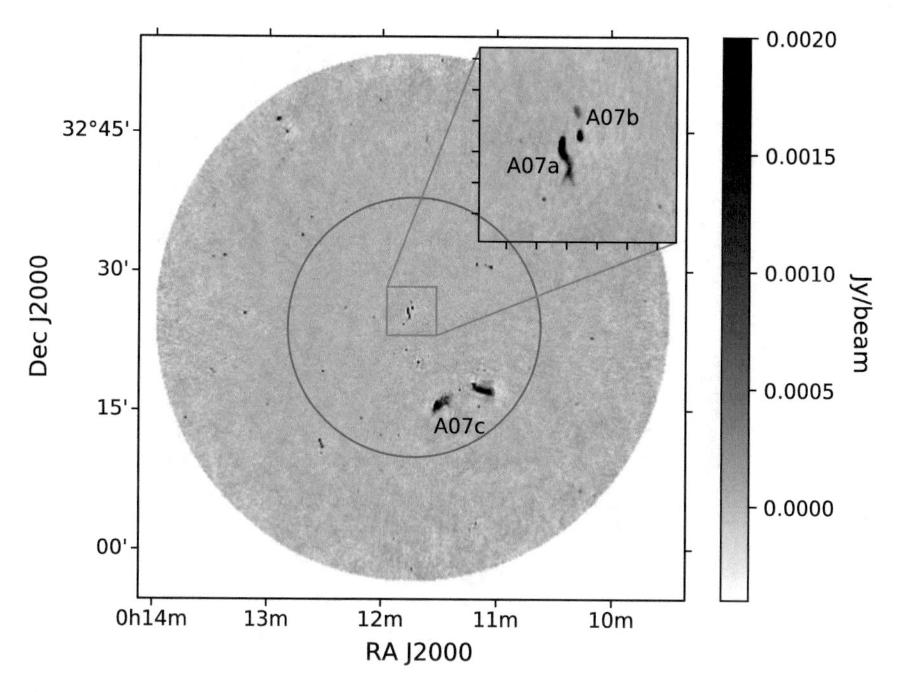

Fig. 4.1 Greyscale image of A07 generated by SPAM. The image has a resolution of 5 arcsec and an rms noise of $\sigma_{\rm rms} = 35$ μJy/beam in quiet region. Near the bright source A07c the noise rises to $\sigma_{\rm rms} = 50$ μJy/beam. The blue circle marks the virial radius of the cluster. An inset in the top right hand corner shows the centre two sources. Three sources are labelled in black in the image. These are A07a, a head tail galaxy, A07b an apparent FRII and A07c a larger FRII radio galaxy near the outskirts of the cluster

The first stage performs standard initial flux and phase calibration using the calibrator sources. The next stage is a self calibration stage. Three rounds of selfcal are performed on the target. The final stage is the direction dependent calibration where a number of bright compact sources in the target field are used to generate a model of the ionosphere. The model is used to find the direction dependent gain solutions for the target field. Figures 4.1, 4.2 and 4.3 show the primary beam corrected image generated by SPAM for each of the clusters in this chapter. Table 4.3 lists the resolution and rms noise of these images.

As I am interested in detecting any large scale structure in these clusters it is necessary to remove any confusing sources from each cluster centre. To do this each field was reimaged with inner uvcut of 4 kλ and Briggs weighting with robust = 0. This produced a high resolution image of each clusters. The inner uvcut ensures that I do not remove any extended emission. For the clusters in this chapter the uvcut used should exclude emission on scales larger than approximately 100 kpc. This image was used to create a model of the point sources in the centre of each cluster which were then subtracted. It should be noted that sources were not subtracted from the

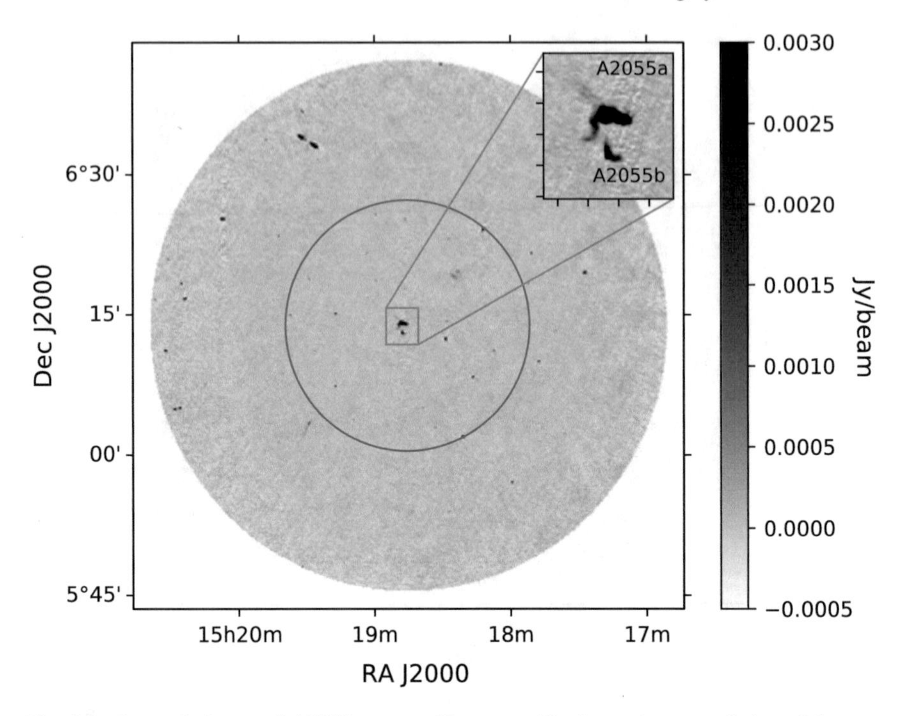

Fig. 4.2 Greyscale image of A2055 generated by SPAM. The image has a resolution of 5 arcsec and an rms noise of $\sigma_{\rm rms} = 50\ \mu$Jy/beam in quiet regions. Near the bright central source the noise rises to $\sigma_{\rm rms} = 0.1$ mJy/beam. The blue circle marks the virial radius of the cluster. An inset in the top right hand corner shows the central two sources

entire image but just the central region of each cluster. Each dataset was imaged a final time using natural weighting, multiscale clean and a uvtaper of 4 kλ. Figure 4.4 shows the primary beam corrected tapered images for A07. Figure 4.5 shows the primary beam corrected tapered images for A2055. Figure 4.6 shows the primary beam corrected tapered images for A1235. Table 4.3 lists the resolution and rms noise of these images. In A2055 and A1235 there is significant emission remaining at the centre of each cluster associated with the diffuse components of radio galaxies in these clustes. This diffuse emission was not captured in the high resolution model.

4.3 Results

4.3.1 A07

Figure 4.1 shows the high resolution full field of view for A07. A blue circle marks the virial radius of the cluster. A head tail radio galaxy (A07a) is located at the centre of the cluster along with what appears to be the the lobes of an FRII (A07b). A large FRII radio galaxy is visible at the outskirts of the cluster.

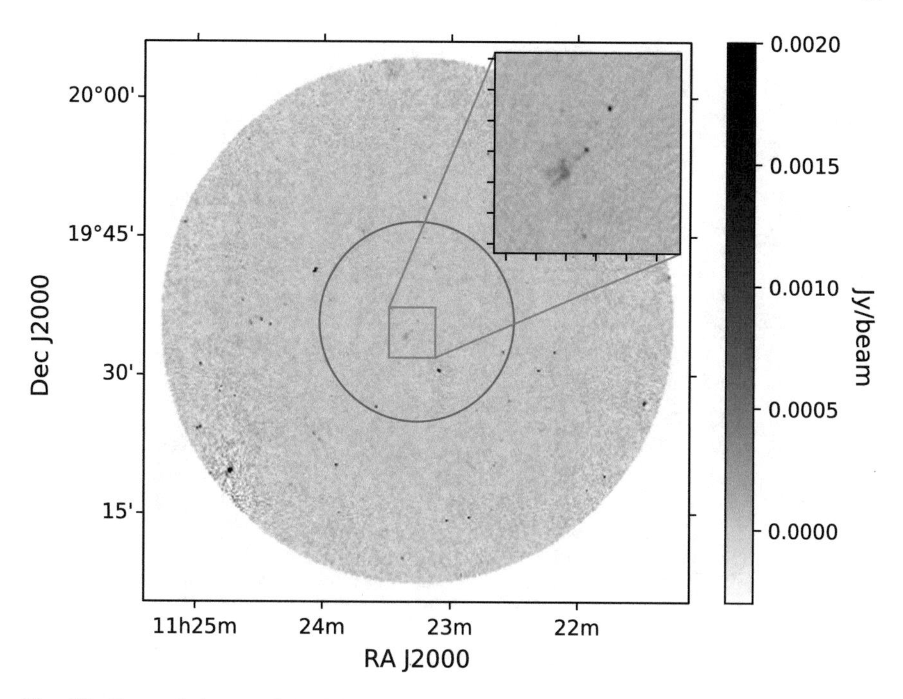

Fig. 4.3 Greyscale image of A1235 generated by SPAM. The image has a resolution of 5 arcsec and an rms noise of $\sigma_{\mathrm{rms}} = 30$ μJy/beam. The blue circle marks the virial radius of the cluster. An inset in the top right hand corner shows the centre source

A07a and A07b are coincident with the centre of the galaxy cluster. Figure 4.7 shows the SDSS g band image of the central region of A07 with high resolution contours overlaid. A galaxy, B20009+32 (SDSS J001146.28 + 322553.7) is coincident with the core of A07a. SDSS spectra show the source has a redshift of $z = 0.098$ which places the source in the cluster. There is no obvious optical source between the proposed lobes of A07b. This would suggest that the host is a high redshift galaxy. There is an optical source, SDSS J001144.01+322553.4, coincident with the proposed southern lobe region of A07b. If SDSS J001144.01 + 322553.4 is the optical counterpart of A07b then the observed radio emission is not that of two lobes of an FRII, but of the unresolved core/jet and a single lobe. SDSS J001144.01 + 322553.4 has a photometric redshift of $z = 0.105 \pm 0.009$ which would place A07b in the cluster.

Figure 4.4 shows the point source subtracted low resolution image of A07. Some emission from the tail of A07a is still visible in the image. Strong negative wells can be seen around A07c suggesting missing flux and limiting the noise and image fidelity. At the centre of the cluster there is a negative well, as well as some emission that does not appear to be associated with any point sources. However due to the limited image fidelity I cannot conclusively say whether this is an imaging artifact or an indication of diffuse emission in A07.

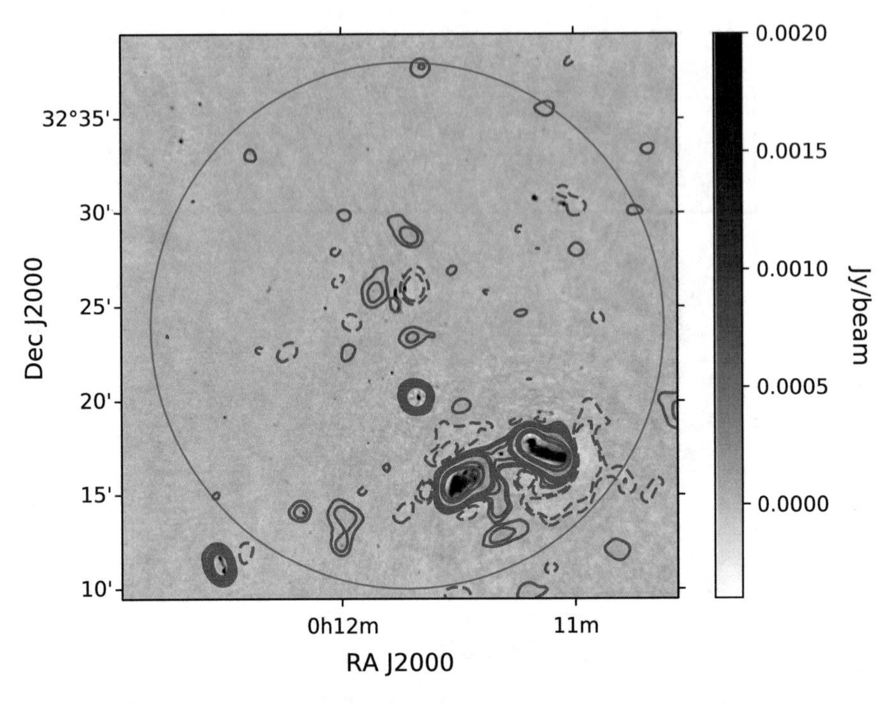

Fig. 4.4 Greyscale image of A07 generated by SPAM. The image has a resolution of 5 arcsec and an rms noise of $\sigma_{\rm rms} = 35$ μJy/beam in quiet region. Near the bright source A07c the noise rises to $\sigma_{\rm rms} = 50$ μJy/beam. The blue circle marks the virial radius of the cluster. Red contours show the point source subtracted low resolution image of A07. The resolution of the low resolution image is 56×55 arcsec. The noise near the centre of the cluster is $\sigma_{\rm rms} = 0.55$ mJy/beam. Near the bright source, A07c, the noise is $\sigma_{\rm rms} = 0.7$ mJy/beam. Contours are at −5, −3, 3, 5, 10, 15, 20, 60, 100 $\times\sigma_{\rm rms} = 0.55$ mJy/beam

4.3.1.1 A07c

An FRII radio galaxy (A07c) with an angular extent of 457 arcsec can be seen coincident with the outskirts of the cluster. This FRII has a large angular extent and was identified as a GRG candidate by Saikia et al. (2002). Figure 4.8 shows the SDSS image of the core region of A07c with high resolution radio contours overlaid. A pair of galaxies SDSS J001119.35 + 321713.8 and SDSS J001119.75 + 321709.0 are coincident with the radio core. SDSS spectra give the redshifts as $z = 0.107$ and $z = 0.106$ respectively. This would place the radio galaxy in the cluster with a projected size of 895 kpc. Such a large projected physical size makes A07c a borderline GRG.

A07 has an FRII morphology with bright hotspots, lobes and only a faint main jet. The western lobe of A07c bends almost 90 degrees at the hotspot. The lobe maintains a fairly constant width until the termination point ∼148 arcsec (289 kpc) from the hotspot. Near the end of the lobe there are two low surface brightness wings,

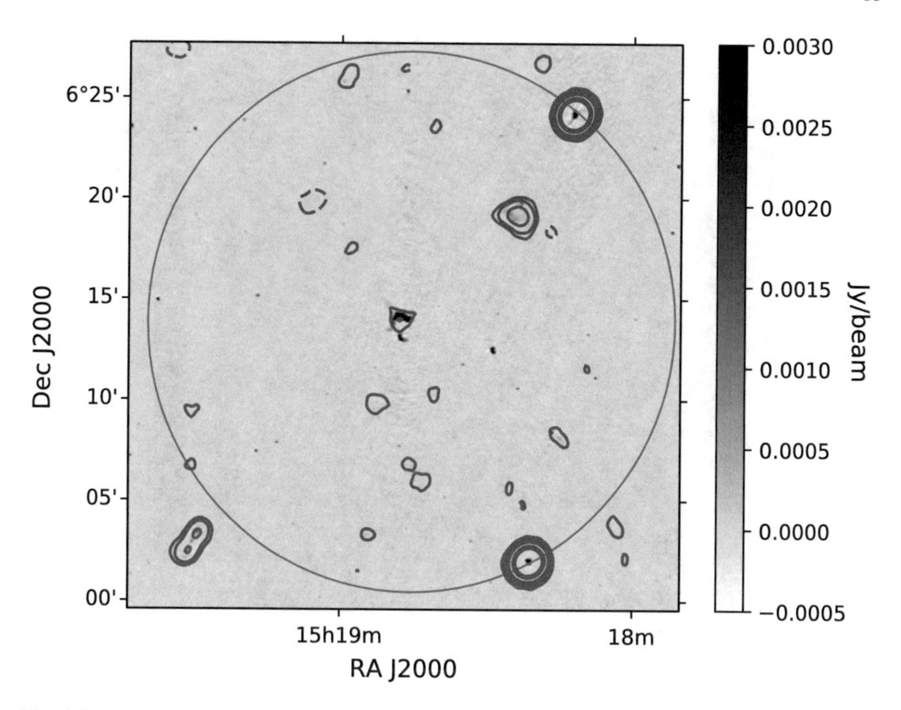

Fig. 4.5 Greyscale image of A2055 generated by SPAM. The image has a resolution of 5 arcsec and an rms noise of $\sigma_{\rm rms} = 50$ μJy/beam in quiet region. Near the bright central source the noise rises to $\sigma_{\rm rms} = 0.1$ mJy/beam. The blue circle marks the virial radius of the cluster. Red contours show the low resolution image of A2055 with point sources subtracted from the centre of the cluster. The resolution of the low resolution image is 59 × 55 arcsec. The noise near the centre of the cluster is $\sigma_{\rm rms} = 0.5$ mJy/beam. Contours are at −5, −3, 3, 5, 10, 15, 20, 60, 100 $\times\sigma_{\rm rms} = 0.55$ mJy/beam. There is residual flux associated with A2055a

roughly perpendicular to the lobe. The eastern lobe does not continue far past the hotspot but instead seems to flow back towards the core. In the low resolution image there is a faint extension to the south of the eastern lobe.

In order to investigate the spectral properties of this source I reimaged the source at the resolution of the TIFR GMRT Sky Survey (TGSS) (Intema et al. 2017) and made a spectral index map from 150–610 MHz. I used a flux cutoff of $7\sigma_{\rm rms}$ where $\sigma_{\rm rms} = 3.5$ mJy/beam is the noise in the TGSS image. Figure 4.9 shows the spectral index map. Both hotspots have spectral indices of $\alpha \sim -0.5$, as would be expected for a region of reacceleration. The spectral index of the eastern lobe increases away from the hotspot, towards to core to $\alpha \sim -1$. The southern spur has a very steep spectral index of $\alpha \sim -2$. The spectral index of the western lobe does not increase with distance from the hotspot. The core of the tube like structure has spectral index of $\alpha \sim -0.8$ which steepens towards the edges. The region of the wings has a steep spectral index of $\alpha \sim -1.5$. Figure 4.10 shows various spectral index slices through A07c. It is clear from Fig. 4.10a that there is little variation in the profile of the spectral index transverse to the direction of the western lobe. Figure 4.10b confirms

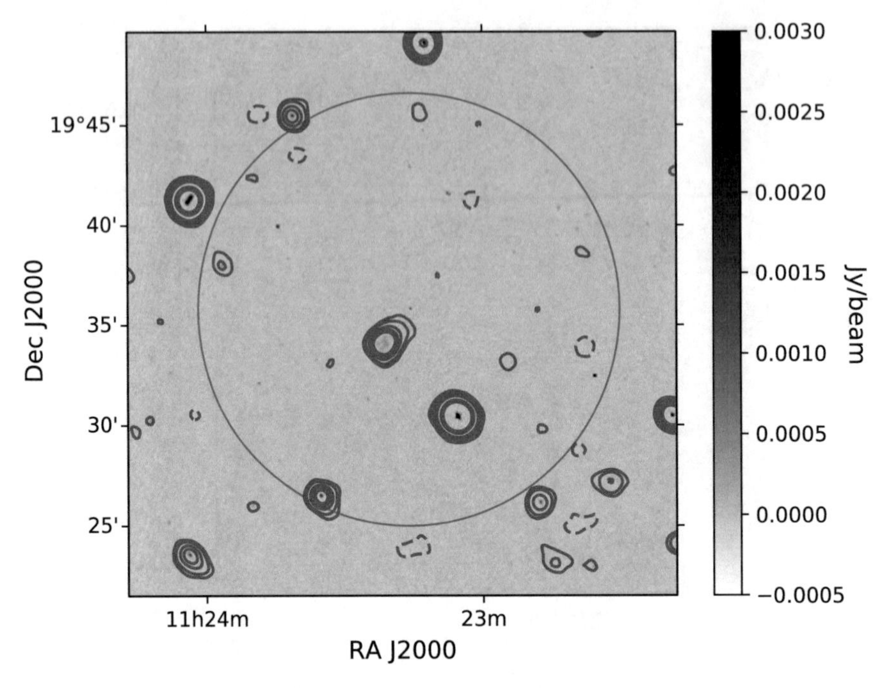

Fig. 4.6 Greyscale image of A1235 generated by SPAM. The image has a resolution of 5 arcsec and an rms noise of $\sigma_{\rm rms} = 30$ μJy/beam. The blue circle marks the virial radius of the cluster. Red contours low resolution image of A1235 with point sources subtracted from the centre of the cluster. The resolution of the low resolution image is 59 × 56 arcsec. The noise near the centre of the cluster is $\sigma_{\rm rms} = 0.25$ mJy/beam. Contours are at $-5, -3, 3, 5, 10, 15, 20, 60, 100 \times \sigma_{\rm rms} = 0.25$ mJy/beam. There is significant residual emsission associated with the diffuse component of A1235a which was not captured in the high resolution model

that while the spectral index in the eastern lobe decreases steadily away from the hotspot the spectral index of the western lobe does not decrease significantly beyond the hotspot.

4.3.2 A1235

Figure 4.3 shows the high resolution image of A1235. There is an unusual radio galaxy in the centre of the cluster with a tuning fork like morphology (A1235a). This source is likely a head tail radio galaxy. Figure 4.11 shows the SDSS image of the field as well as the high resolution GMRT image of the tuning fork radio galaxy. The optical source SDSS J112318.41+193441.3 is coincident with the core of A1235a with a spectroscopic redshift of $z = 0.104$ placing the source in the cluster.

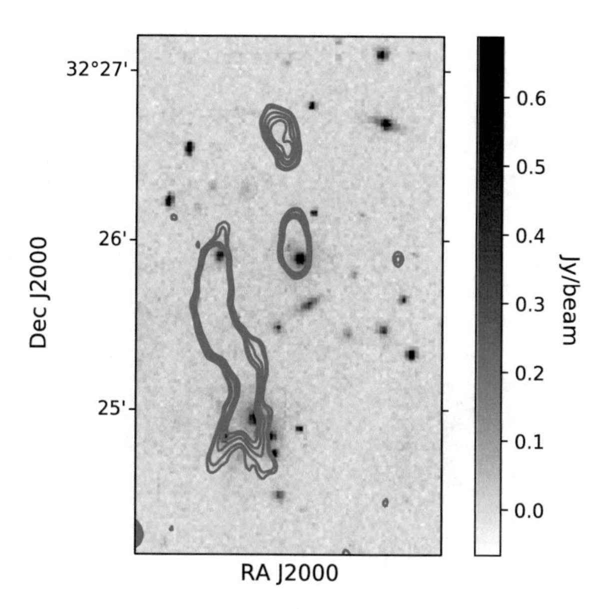

Fig. 4.7 SDSS g band image of central sources in A07. High resolution GMRT data are overlaid in blue contours. Contours are at 3, 5, 10, 15, 20 $\times\sigma_{\rm rms}$ where $\sigma_{\rm rms} = 35$ μJy/beam is the rms noise in the image

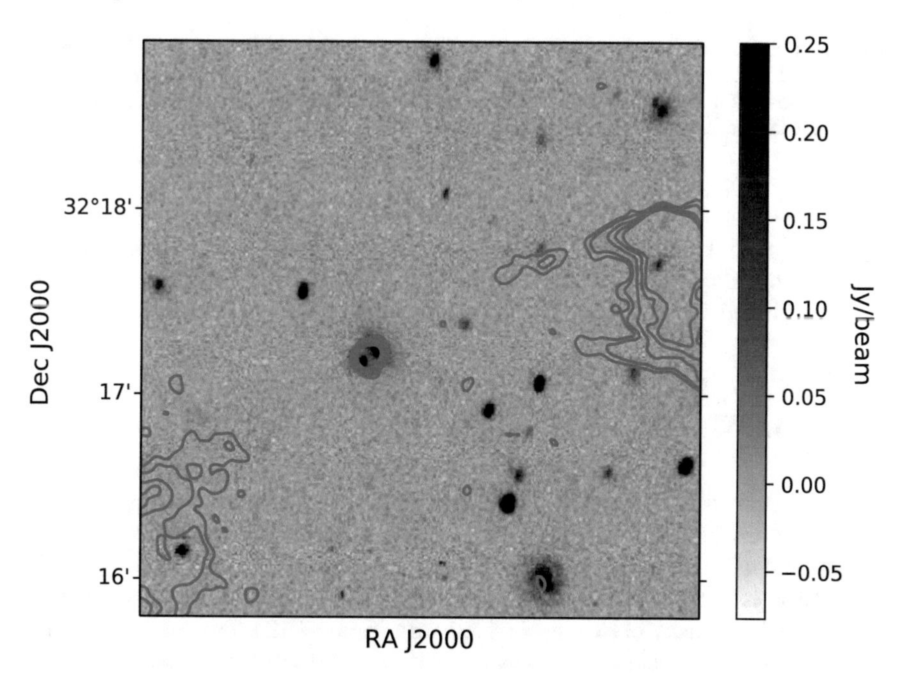

Fig. 4.8 SDSS g band image of the core of radio galaxy A07c. High resolution GMRT data are overlaid in blue contours. Contours are at 3, 5, 10, 15, 20 $\times\sigma_{\rm rms}$ where $\sigma_{\rm rms} = 35$ μJy/beam is the rms noise in the image

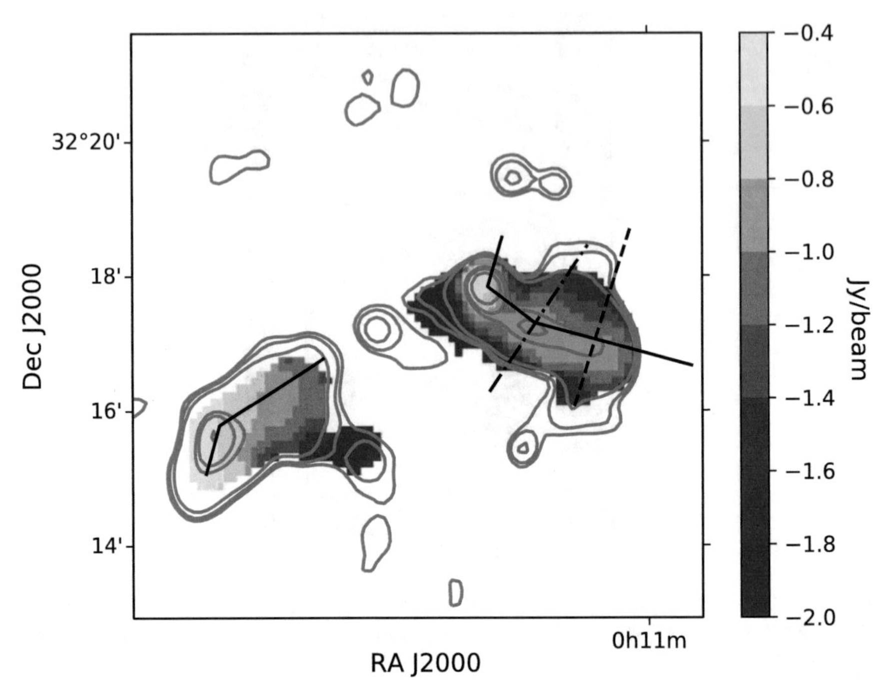

Fig. 4.9 Spectral index map of A07c. Contours for the GMRT data reimaged to the TGSS beam are overlaid in grey. Contours are at 3,5,10,80,100,150 $\times\sigma_{\rm rms}$ where $\sigma_{\rm rms} = 0.4$ mJy/beam is the noise in the GMRT image. Solid black lines show the spectral index slices for the east and west lobe while the dahed black lines show the transverse slices for the west lobe

Figure 4.3 shows the point source subtracted low resolution image of A1235. The tail of the central radio galaxy is still clearly visible suggesting significant emission on scales less than <4 kλ. There is no evidence for any large scale diffuse emission in A1235 based on these data.

4.3.3 A2055

Figure 4.2 shows the high resolution image of A2055. There are two sources seen at the centre of A2055, labelled here as A2055a and A2055b. A2055a is a very bright head tail radio galaxy with an integrated flux density of $S_{610\,\rm MHz} = 800 \pm 80$ mJy. There is a low surface brightness extension to the north east of the tail with an integrated flux density of $S_{610\,\rm MHz} = 4 \pm 1$ mJy. Figure 4.12 shows the SDSS image of the field. The optical source SDSS J151845.73 + 061356.1 is coincident with the core of A2055a and has a spectroscopic redshift of $z = 0.102$, making A2055a a cluster member.

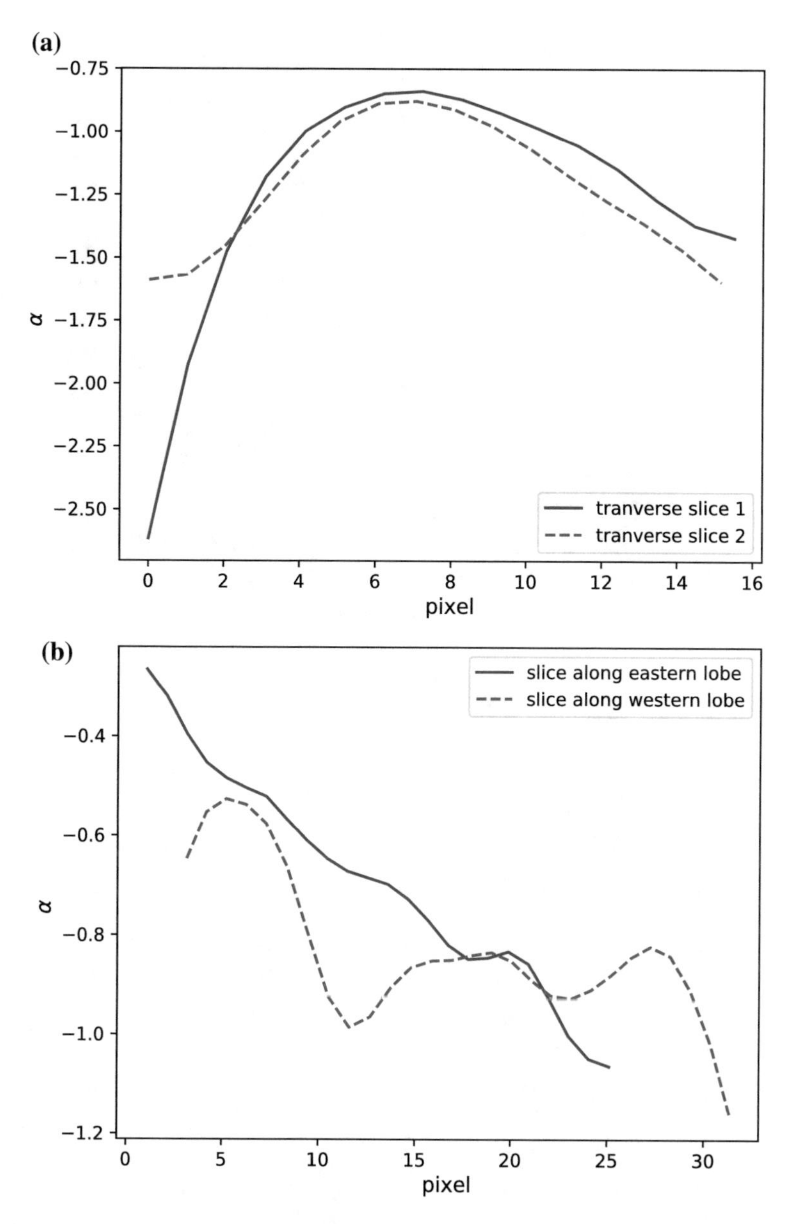

Fig. 4.10 **a** spectral index slices across the western lobes of A07c. **b** spectral index slices along the lobes of A07c

A2055b is a wide angle tail galaxy with an integrated flux density of $S_{610\,\mathrm{MHz}} = 74 \pm 7$ mJy. Figure 4.12 shows that the optical source SDSS J151847.71+061258.9 is coincident with the core of A2055b with a spectroscopic redshift of $z = 0.499$ placing A2055b behind the cluster.

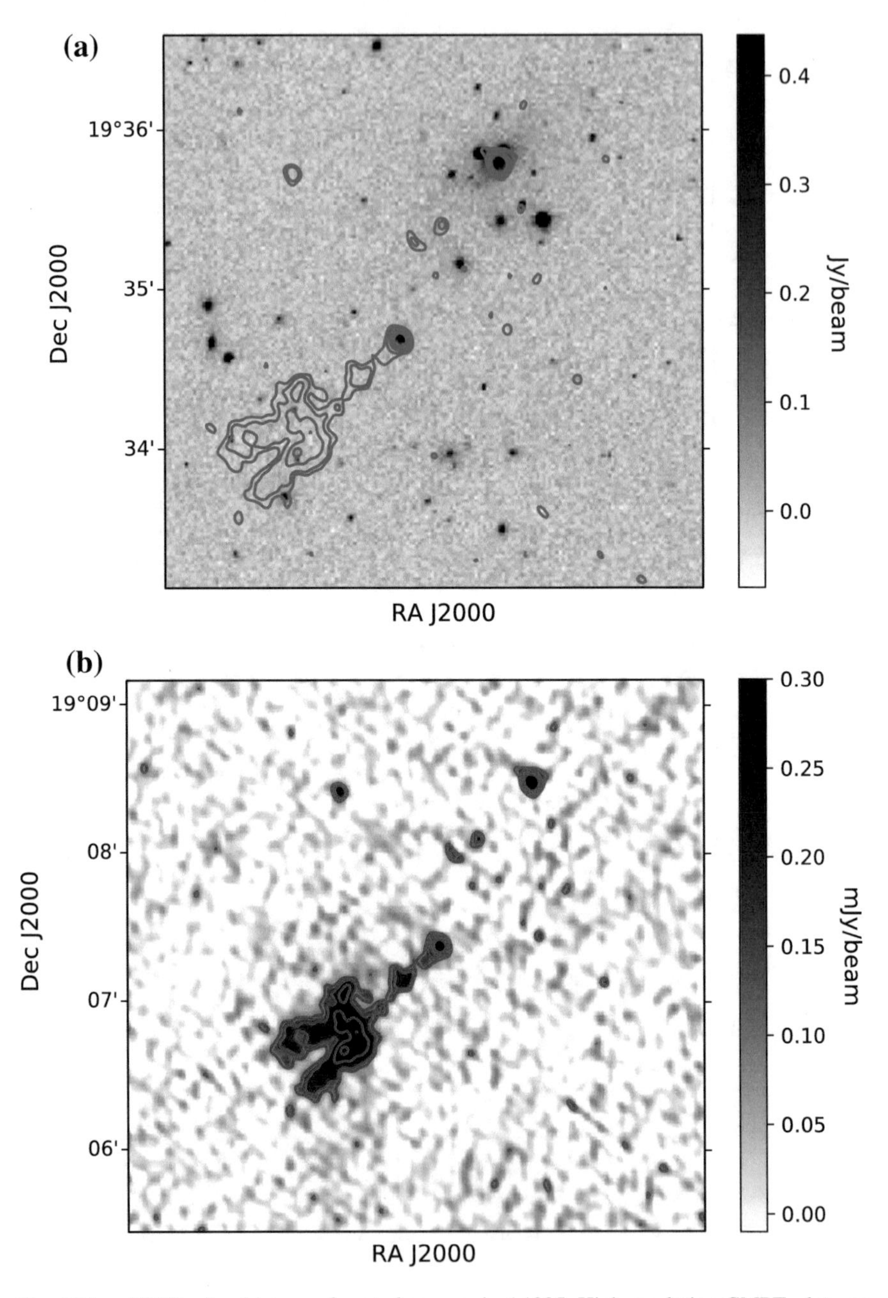

Fig. 4.11 **a** SDSS g band image of central sources in A1235. High resolution GMRT data are overlaid in blue contours. Contours are at 3, 5, 10, 15, 20 $\times\sigma_{\rm rms}$ where $\sigma_{\rm rms} = 35$ μJy/beam is the rms noise in the image. **b** Greyscale image of A1235 generated by SPAM. The image has a resolution of 5 arcsec and an rms noise of $\sigma_{\rm rms} = 30$ μJy/beam. Contours are at 3, 5, 10, 15, 20 $\times\sigma_{\rm rms}$

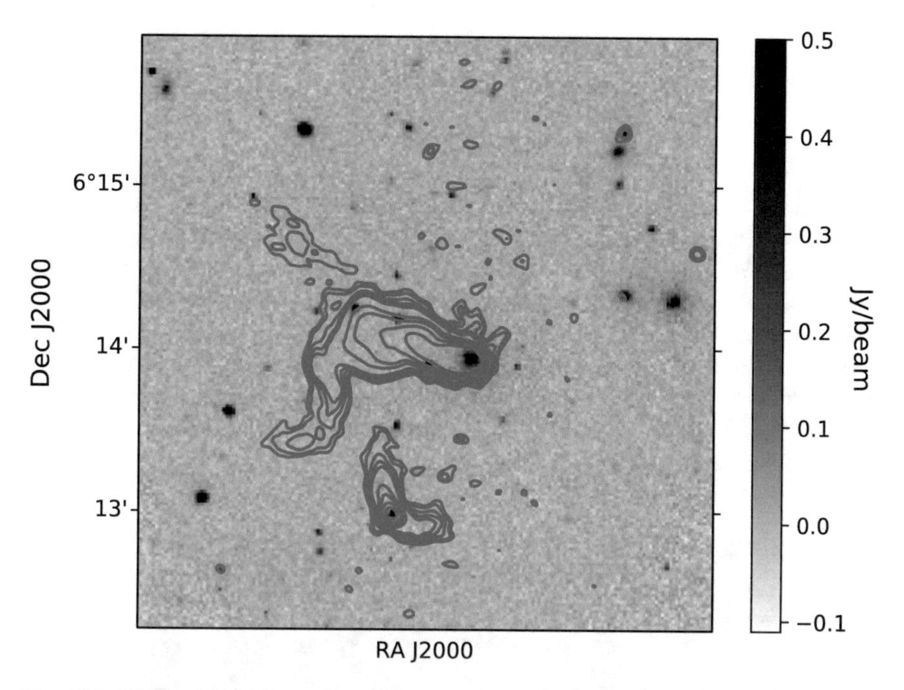

Fig. 4.12 SDSS g band image of central sources in A2055. High resolution GMRT data are overlaind in blue contours. Contours are at 3, 5, 10, 15, 20, 60, 100, 200, 400 $\times\sigma_{\rm rms}$ where $\sigma_{\rm rms} = 70$ μJy/beam is the noise near A2055a

Figure 4.5 shows the point source subtracted low resolution image of A2055. There is still some residual emission associated with A2055a however there is no evidence for large scale diffuse emission in A2055 based in these data.

4.3.4 New Upper Limits

There was no detection of large-scale diffuse emission in any of the clusters. In order to place an upper limit on the power of any potential radio halo in these clusters I follow the procedure in Johnston-Hollitt and Pratley (2017) and Bonafede et al. (2017). I generate a synthetic halo in an empty region of the map with a brightness profile given by

$$I(r) = I_0 e^{-2.6r} \tag{4.1}$$

where r is the normalised distance from the centre of the halo (Murgia et al. 2009). I_0 is defined as

$$I_0 = \frac{S_\nu}{\int_0^{2\pi} d\theta \int_0^{R_{\rm H}} r I(r) dr} \tag{4.2}$$

where R_H is the radius of the halo in arcsec S_ν is the integrated flux density of the halo. The brightness in a pixel is then given by

$$B(r) = I_0 e^{-2.6r} \Delta A \quad \text{Jy pixel} \tag{4.3}$$

where ΔA is the area of a pixel in arcsec^2. The model halo is then Fourier transformed into the uv plane and added to the measured visibilities and imaged. As in Johnston-Hollitt and Pratley (2017) I consider the detection threshold to be when 25% of the simulated halo surface area is recovered. If more than 25% of the surface area of the halo is recovered the halo is considered detected. In this case I generate a new synthetic halo with a lower integrated flux density. If less than 25% of the surface area of the halo is recovered I generate a new synthetic halo with a higher integrated flux density. Figure 4.13, 4.14 and 4.15 show the simulated halos at the detection threshold for each cluster.

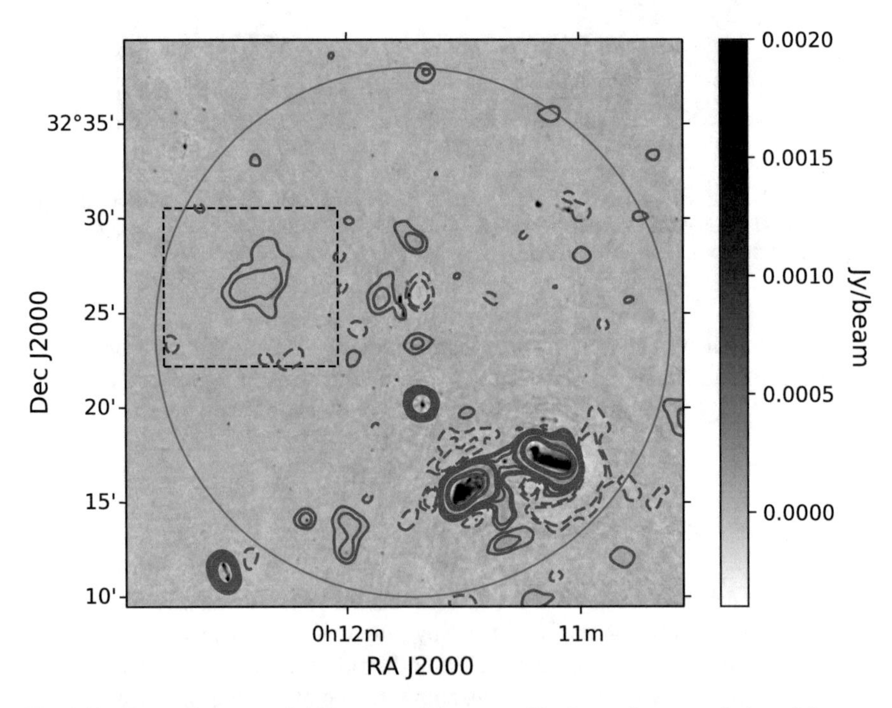

Fig. 4.13 Greyscale image of A07 generated by SPAM. The image has a resolution of 5 arcsec and an rms noise of $\sigma_{\rm rms} = 35$ μJy/beam in quiet region. Near the bright source A07c the noise rises to $\sigma_{\rm rms} = 50$ μJy/beam. The blue circle marks the virial radius of the cluster. Red contours show the point source subtracted low resolution image of the A07 data with an injected radio halo. The simulated radio halo shown in this image is at the detection threshold described in Sect. 4.3.4. The black rectangle shows the region in which the simulated halo was placed. The resolution of the low resolution image is 56 × 55 arcsec. The noise near the centre of the cluster is $\sigma_{\rm rms} = 0.55$ mJy/beam. Near the bright source, A07c, the noise is $\sigma_{\rm rms} = 0.7$ mJy/beam. Contours are at −5, −3, 3, 5, 10, 15, 20, 60, 100 $\times\sigma_{\rm rms} = 0.55$ mJy/beam

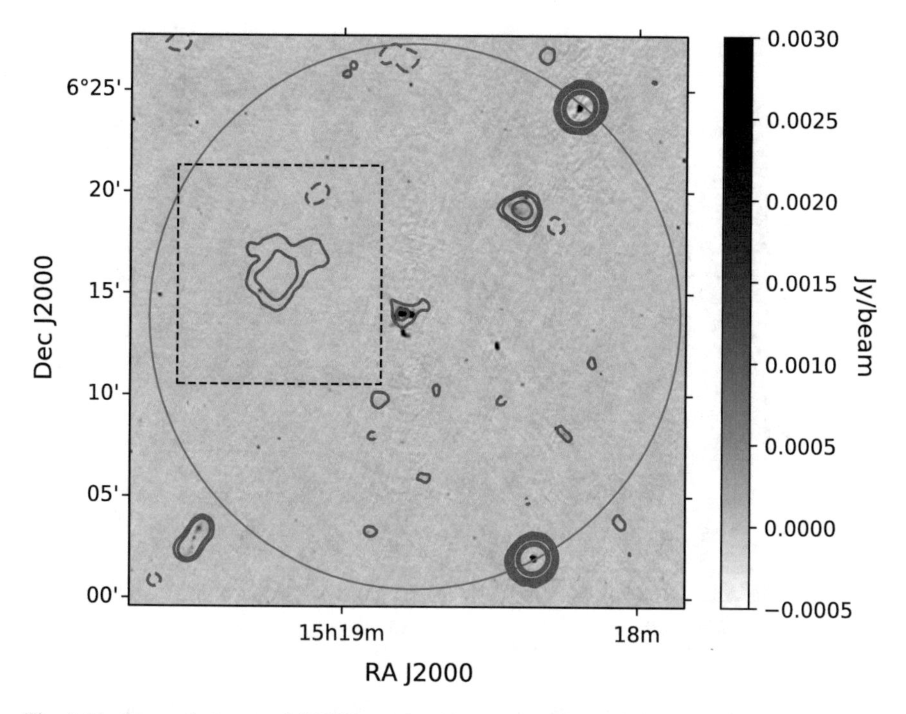

Fig. 4.14 Greyscale image of A2055 generated by SPAM. The image has a resolution of 5 arcsec and an rms noise of $\sigma_{\mathrm{rms}} = 50$ μJy/beam in quiet region. Near the bright central source the noise rises to $\sigma_{\mathrm{rms}} = 0.1$ mJy/beam. The blue circle marks the virial radius of the cluster. Red contours show the point source subtracted low resolution image of the A2055 data with an injected radio halo. The simulated radio halo shown in this image is at the detection threshold described in Sect. 4.3.4. The black rectangle shows the region in which the simulated halo was placed. The resolution of the low resolution image is 59 × 55 arcsec. The noise near the centre of the cluster is $\sigma_{\mathrm{rms}} = 0.5$ mJy/beam. Contours are at −5, −3, 3, 5, 10, 15, 20, 60, 100 $\times \sigma_{\mathrm{rms}} = 0.5$ mJy/beam

Table 4.4 shows the upper limit P_{610} for each cluster and Fig. 4.16 shows the position of these upper limits in relation to previously reported halo detections and limits.

4.4 Discussion

In the previous section I showed that there was no evidence of diffuse emission in any of the clusters observed. The upper limits found for P_{610} are consistent with those found in the GMRT radio halo survey (GRHS) (Venturi et al. 2008) and the Extended GMRT radio halo survey (EGRHS) (Kale et al. 2013). The upper limits found found in A07 and A2055 are toward the high end of the range of upper limits die to bright sources limiting the noise. The upper limits found here are all below the expected power from the $P_{610} - L_X$ but within the observed scatter of the relationship.

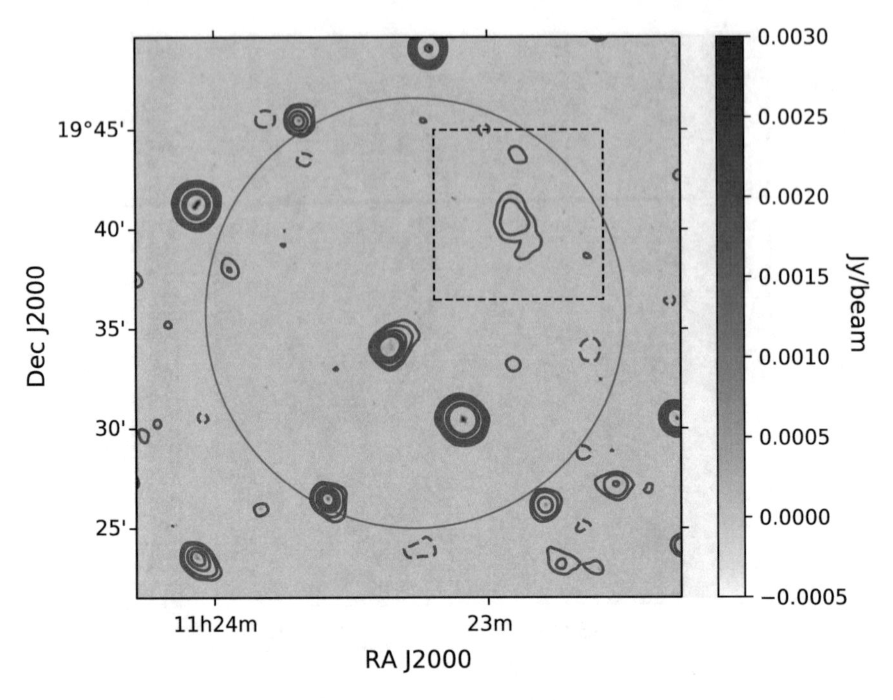

Fig. 4.15 Greyscale image of A1235 generated by SPAM. The image has a resolution of 5 arcsec and an rms noise of $\sigma_{rms} = 30$ µJy/beam. The blue circle marks the virial radius of the cluster. Red contours show the point source subtracted low resolution image of the A1235 data with an injected radio halo. The simulated radio halo shown in this image is at the detection threshold described in Sect. 4.3.4. The black rectangle shows the region in which the simulated halo was placed. The resolution of the low resolution image is 59 × 56 arcsec. The noise near the centre of the cluster is $\sigma_{rms} = 0.25$ mJy/beam. Contours are at −5, −3, 3, 5, 10, 15, 20, 60, 100 $\times\sigma_{rms} = 0.25$ mJy/beam

The clusters are expected to host radio halos as each cluster has $\Gamma < -1.0$ indicating a highly disturbed morphology. It is unusual for merging clusters not to host halos. Cassano et al. (2010) find that out of 16 dynamically disturbed clusters only 4 do not host a radio halo. Cassano et al. (2010) suggest that the non detection of a radio halo in these clusters is due to the low X-ray luminosity ($L_X < 8 \times 10^{44}$ erg s^{-1}) of three of the clusters and the high redshift ($z > 0.32$) of the fourth cluster. Based on the reacceleration model, these clusters would be expected to have a lower break frequency in the radio halo spectra and thus require very low frequency observations to be detected.

It is possible that these clusters host ultra steep spectrum radio halos (USSRHs). USSRHs are radio halos with a spectral index of $\alpha < -2.0$. A521 is the first USSRH discovered with a spectral index of $\alpha \approx -1.9$ and is considered the prototypical USSRH (Brunetti et al. 2008). It is only clearly detected at frequencies below 330 MHz. The most constraining upper limit is for A1235 which requires that $S_{610} < 21$ mJy. An USSRH with $\alpha = -2$ in A1235 would have an integrated spectral index at 150 MHz of $S_{150} < 330$ mJy. Observations with the LOw Frequency ARray

Table 4.4 Upper limit $P_{610\,\mathrm{MHz}}$ for A07, A1235 and A2055. Column 1 lists the cluster names, column 2 shows the rms noise in the low resolution image, column 3 shows the resolution of the image, column 4 shows the upper limit S_{610} and column 5 shows the upper limit P_{610}

Cluster	σ_{rms} (mJy/beam)	Resolution (arcsec × arcsec)	S_{610} (mJy)	P_{610} (Watts Hz^{-1})
A07	0.55	56 × 55	53	1.5×10^{24}
A2055	0.5	59 × 55	67	1.8×10^{24}
A1235	0.25	59 × 56	21	5.8×10^{23}

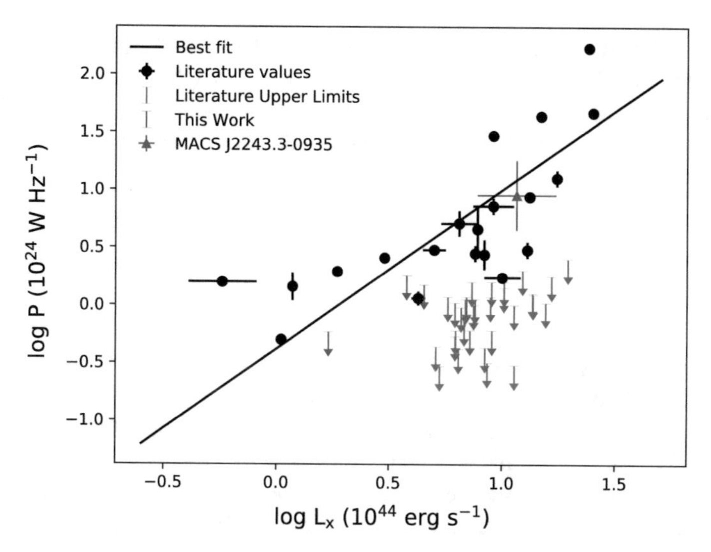

Fig. 4.16 $P_{610} - L_X$ relationship. Black circles shown known radio halo powers taken from Yuan et al. (2015). Blue upper arrows show the upper limits for clusters in the GMRT radio halo survey (Venturi et al. 2008) and the Extended GMRT radio halo survey (Kale et al. 2013). Red arrows show the upper limits calculated for the clusters in this chapter. The black line shows the best fit to the $P_{610} - L_X$ data as reported in Yuan et al. (2015)

(LOFAR) would be able to search for a USSRH in these clusters and confirm whether or not they are truly radio quiet.

The consistently flat spectral index in the western lobe of A07c suggest some form of reacceleration is ongoing in the lobe. There is evidence for extreme bending at the hotspot so one likely scenario is that we are seeing the lobe highly projected. In this scenario the lobe is expanding into its environment towards us. What is see then is the leading front of the lobe, where interaction with the environment accelerates the particles, flattening the spectral index. The steep spectrum wings are the part of the backflowing lobe which is visible behind the leading edge of the lobe.

A07c is located at the edge of the cluster and so an alternative scenario could be that the lobe has been compressed by shocks near the cluster outskirts, accelerating lobe electrons. Such reacceleration has been reported in radio galaxies before. Ensslin et al. (2001) suggest that a powerful shock wave is responsible for the flat spectrum west lobe in NGC 315 as well as the the source of the extreme bend in the western lobe.

4.5 Conclusion and Future Work

I have placed upper limits on the radio power at 610 MHz for three clusters, A07, A1235 and A2055. These limits are below the $P_{610} - L_X$ and rule out bright radio halo in these clusters. I have identified these clusters as potential hosts for USSRH. Observations with LOFAR should be capable of confirming whether or not these clusters host USSRH.

4.5.1 Future Work

I have begun an observing campaign with the VLA of 10 galaxy clusters at S band in C and D configuration. The purpose of this campaign is to identify new halos and determine the range of Γ where detecting a radio halo is most likely. The criteria used for selecting this sample was that $z > 0.13$ so that the VLA in D-config would be sensitive to emission as large as 1 Mpc. I also required that there be no bright central radio galaxies in order to avoid confusion when identifying radio halos.

References

Böhringer H, Voges W, Huchra JP, McLean B, Giacconi R, Rosati P, Burg R, Mader J, Schuecker P, Simiç D, Komossa S, Reiprich TH, Retzlaff J, Trümper J (2000) The northern ROSAT All-Sky (NORAS) galaxy cluster survey. I. X-Ray properties of clusters detected as extended X-Ray sources. APJS 129:435–474. https://doi.org/10.1086/313427, arXiv:astro-ph/0003219

Bonafede A, Cassano R, Brüggen M, Ogrean GA, Riseley CJ, Cuciti V, de Gasperin F, Golovich N, Kale R, Venturi T, van Weeren RJ, Wik DR, Wittman D (2017) On the absence of radio haloes in clusters with double relics. MNRAS 470:3465–3475. https://doi.org/10.1093/mnras/stx1475, arXiv:1706.04203

Brunetti G (2016) The challenge of turbulent acceleration of relativistic particles in the intra-cluster medium. Plasma Phys. Control. Fus. 58(1):014011. https://doi.org/10.1088/0741-3335/58/1/014011, arXiv:1509.03299

Brunetti G, Giacintucci S, Cassano R, Lane W, Dallacasa D, Venturi T, Kassim NE, Setti G, Cotton WD, Markevitch M (2008) A low-frequency radio halo associated with a cluster of galaxies. Nature 455:944–947. https://doi.org/10.1038/nature07379, arXiv:0810.4288

Brunetti G, Jones TW (2014) Cosmic rays in galaxy clusters and their nonthermal emission. Int. J. Modern Phys. D 23(04):1430007. https://doi.org/10.1142/S0218271814300079, http://arxiv.org/abs/1401.7519, http://www.arxiv.org/pdf/1401.7519.pdf

Cassano R (2010) The radio-X-ray luminosity correlation of radio halos at low radio frequency. Application of the turbulent re-acceleration model. A&A 517:A10, https://doi.org/10.1051/0004-6361/200913622, arXiv:1004.1171

Cassano R, Brunetti G, Norris RP, Röttgering HJA, Johnston-Hollitt M, Trasatti M (2012) Radio halos in future surveys in the radio continuum. A&A 548:A100. https://doi.org/10.1051/0004-6361/201220018, arXiv:1210.1020

Cassano R, Ettori S, Giacintucci S, Brunetti G, Markevitch M, Venturi T, Gitti M (2010) On the connection between giant radio halos and cluster mergers. ApJ721:L82–L85, https://doi.org/10.1088/2041-8205/721/2/L82, arXiv:1008.3624

Ensslin TA, Simon P, Biermann PL, Klein U, Kohlec S, Kronberg PP, Mack KH (2001) Signatures in a giant radio galaxy of a cosmological shock wave at intersecting filaments of galaxies. ApJ 549:L39–L42. https://doi.org/10.1086/319131, arXiv:astro-ph/0012404

Feretti L, Giovannini G, Govoni F, Murgia M (2012) Clusters of galaxies: observational properties of the diffuse radio emission. Astron Astrophys Rev 20(1):54, https://doi.org/10.1007/s00159-012-0054-z, https://doi.org/10.1007/s00159-012-0054-z, arXiv:1205.1919v1

Intema HT, Jagannathan P, Mooley KP, Frail DA (2017) The GMRT 150 MHz all-sky radio survey. First alternative data release TGSS ADR1. A&A 598:A78, https://doi.org/10.1051/0004-6361/201628536, arXiv:1603.04368

Intema HT (2014) SPAM: A data reduction recipe for high-resolution, low-frequency radio-interferometric observations. Astron Soc India Conf Ser Astron Soc India Conf Ser 13(1402):4889

Johnston-Hollitt M, Pratley L (2017) Upper limits on a radio halo in Abell 3667 at 1.4 GHz. ArXiv e-prints 1706.04930

Kale R, Venturi T, Giacintucci S, Dallacasa D, Cassano R, Brunetti G, Macario G, Athreya R (2013) The extended GMRT radio halo survey. I. New upper limits on radio halos and mini-halos. A&A 557:A99, https://doi.org/10.1051/0004-6361/201321515, arXiv:1306.3102

Murgia M, Govoni F, Markevitch M, Feretti L, Giovannini G, Taylor GB, Carretti E (2009) Comparative analysis of the diffuse radio emission in the galaxy clusters A1835, A2029, and Ophiuchus. A&A 499:679–695. https://doi.org/10.1051/0004-6361/200911659, arXiv:0901.1943

Piffaretti R, Arnaud M, Pratt GW, Pointecouteau E, Melin JB (2011) The MCXC: a meta-catalogue of x-ray detected clusters of galaxies. A&A 534:A109. https://doi.org/10.1051/0004-6361/201015377, arXiv:1007.1916

Sadat R, Blanchard A, Kneib JP, Mathez G, Madore B, Mazzarella JM (2004) Introducing BAX: A database for X-ray clusters and groups of galaxies. A&A 424:1097–1100. https://doi.org/10.1051/0004-6361:20034029, arXiv:astro-ph/0405457

Saikia DJ, Thomasson P, Spencer RE, Mantovani F, Salter CJ, Jeyakumar S (2002) CSSs in a sample of B2 radio sources of intermediate strength. A&A 391:149–157. https://doi.org/10.1051/0004-6361:20020807, arXiv:astro-ph/0206049

Venturi T, Giacintucci S, Dallacasa D, Cassano R, Brunetti G, Bardelli S, Setti G (2008) GMRT radio halo survey in galaxy clusters at z = 0.2-0.4. II. The eBCS clusters and analysis of the complete sample. A&A 484:327–340. https://doi.org/10.1051/0004-6361:200809622, arXiv:0803.4084

Wen ZL, Han JL (2013) Substructure and dynamical state of 2092 rich clusters of galaxies derived from photometric data. MNRAS 436:275–293. https://doi.org/10.1093/mnras/stt1581, arXiv:1307.0568

Yuan ZS, Han JL, Wen ZL (2015) The scaling relations and the fundamental plane for radio halos and relics of galaxy clusters. ApJ813(1):77, http://stacks.iop.org/0004-637X/813/i=1/a=77

Chapter 5
LOFAR HBA Observations of NGC 6251

5.1 Introduction

In this chapter I present total intensity and polarised intensity observations of the nearby giant radio galaxy NGC 6251 with LOFAR HBA. NGC 6251 is a giant radio galaxy with a borderline FRI/FRII morphology. The main jet and lobe are centre brightened like an FRI however there is a hot spot or 'warm spot' in the lobe suggestive of an FRII. In contrast the counter jet and lobe are edge brightened with another hot spot in the southern lobe. The radio power at 178 MHz is $P_{178\mathrm{MHz}} \approx 1.4 \times 10^{25}$ W Hz^{-1} (Waggett et al. 1977) which is near the 178 MHz boundary between FRI and FRII, $P_{178\mathrm{MHz}} \approx 2 \times 10^{25} h_{50}^{-2}$ W Hz^{-1} or $P_{178\mathrm{MHz}} \approx 1 \times 10^{25} h_{70}^{-2}$ W Hz^{-1} (Fanaroff and Riley 1974). The images presented in this chapter are the highest sensitivity and resolution images of NGC 6251 at these frequencies to date. I will also present the first detailed spectral index maps of the source at such low frequencies.

Joint analysis of the X-ray and radio data show that if equipartition is assumed then the northern jet of NGC 6251 can only be confined if no beaming or projection effects are present (Evans et al. 2005). Evans et al. (2005) show that when these effects are included, equipartition assumptions do not provide the pressure necessary to balance the internal pressure of the jet with the external pressure. This would suggest that some deviation of the jet energetics from equipartition is needed to achieve pressure balance.

There have been many radio observations of NGC 6251. The first observations were carried out by Waggett et al. (1977) at 1.4 GHz and 150 MHz. Perley et al. (1984) present detailed high resolution Very Large Array (VLA) observations of the main jet in NGC 6251 at 1.4 GHz. Mack et al. (1997, 1998) present observations of the large scale structure of NGC 6251 from 325 MHz to 10 GHz. Observations also suggest that NGC 6251 is highly polarised, as much as 70% in some regions (Willis et al. 1978; Stoffel and Wielebinski 1978; Saunders et al. 1981; Mack et al. 1997; Perley et al. 1984).

There have been high energy observations of NGC 6251. Fermi detected NGC 6251 as 1FGL J1635.4+8228 in the first-year FERMI catalogue (Abdo et al. 2010)

T. Cantwell, *Low Frequency Radio Observations of Galaxy Clusters and Groups*,
Springer Theses, https://doi.org/10.1007/978-3-319-97976-2_5

Table 5.1 Observation details of NGC6251

	HBA
Project code	LC0_012
Date	23-Aug-2013
Central frequency (MHz)	150
Time on target (hrs)	6.5
Usable time (hrs)	6.5
Bandwidth (MHz)	80
Usable bandwidth (MHz)	63
No. Channels/SB	64
No. Averaged channels/SB	4
% flagged	38%
Sensitivity	2 mJy/beam
Angular resolution	40 arcsec
FOV	$\sim$6 deg^2

and as 2FGL J1629.4+8236 in the Fermi second-year catalogue (Nolan et al. 2012). The 95% LAT error on the position of 2FGL J1629.4+8236 includes both the jet and lobe of NGC 6251. Takeuchi et al. (2012) observed NGC 6251 with Suzaku and detected diffuse X-ray emission in the north west lobe of NGC 6251. They argue that 2FGL J1629.4+8236 is consistent with non-thermal inverse Compton emission from the lobes based on detailed modelling of the SED in the lobe.

The aim of the work presented in this chapter is two-fold. The first aim is to investigate the low frequency spectral behaviour of NGC 6251 and re-examine the pressure balance in NGC 6251 taking into account the new HBA data. The second is to probe the environment and source structure using the high resolution Faraday spectra obtained using the LOFAR HBA. In Sect. 5.2 I describe the observations and data reduction. In Sect. 5.3 I present the results which I discuss in Sect. 5.4 before making my concluding remarks in Sect. 5.5.

At a redshift of 0.02471, 1 arcsec corresponds to a physical scale of 0.498 kpc (Wright 2006).

5.2 Observations and Data Reduction

NGC 6251 was observed during LOFAR's cycle 0. For a summary of the observations taken see Table 5.1. Data were taken in interleaved mode, with scans alternating between the target and the flux calibrator. This mode was used in early LOFAR cycles due to uncertainty in the gain stability. The calibrator 3C295 was observed for 2 min and the target scans were 10 min long.

5.2.1 HBA Observations

5.2.1.1 HBA Calibration and Imaging

NGC 6251 was observed by the LOFAR HBA on the 23rd of August 2013. Data were taken in interleaved mode with scans alternating between 10 min scans on the target and 2 min scans on the primary calibrator 3C295. An initial flagging step was performed using AOFLAGGER (Offringa et al. 2012).[1] 3C295 was then calibrated using BlackBoard SelfCal (BBS) and a simple two component model. The flux scale is set using Scaife and Heald (2012). These solutions were transferred to the target and then a phase only self cal was performed on each sub-band using the LOFAR global skymodel (gsm). The data were imaged with AWIMAGER to generate a new skymodel which was then used to perform a single round of phase only selfcal. The data were combined into 18 bands of 3.515 MHz each. The channel width was kept at 48 kHz in order to maintain good maximum Faraday depth.

Final imaging was carried out using AWIMAGER (Tasse et al. 2013). AWIMAGER is an imaging program which utilises the A-projection algorithm (Bhatnagar et al. 2008) in order to correct for direction-dependent effects caused by LOFAR's varying primary beam. 63 MHz of the 80 MHz total bandwidth was used. Each band was imaged separately, with a uvcut of 3k λ, to a resolution of 40 arcsec. At this point the flux scale was corrected as described in Sect. 5.2.1.2. The flux corrected images were combined together using a weighted average. Figure 5.1 shows the full field of view with a resolution of 40 arcsec and an rms of 2 mJy/beam. Figure 5.2 shows a zoomed in image of NGC6251. The expected image sensitivity with these parameters is approximately 0.2 mJy/beam. However due to direction dependent effects which have not been calibrated the noise is about a factor of 10 higher. This increase in noise is typical for data which have not undergone direction dependent calibration (van Weeren et al. 2016b).

5.2.1.2 HBA Flux Scale

There are known problems with the HBA flux scale (Heald et al. 2015; Hardcastle et al. 2016). The LOFAR primary beam is elevation dependent, leading to different primary beams when observing the calibrator source and the target source. This difference should be accounted for when transferring the amplitude gains from the calibrator source to the target field. However as the LOFAR HBA beam is poorly constrained it is not currently possible to include this effect during calibration. This leads to a frequency dependent effect on the HBA fluxes. In order to correct for this effect, I follow the flux bootstrapping procedure outlined in Hardcastle et al. (2016).[2]

[1] AOFLAGGER can be found at https://sourceforge.net/projects/aoflagger.

[2] The code used to fit the flux correction was written by Martin Hardcastle and is available from https://github.com/mhardcastle/surveys-pipeline.

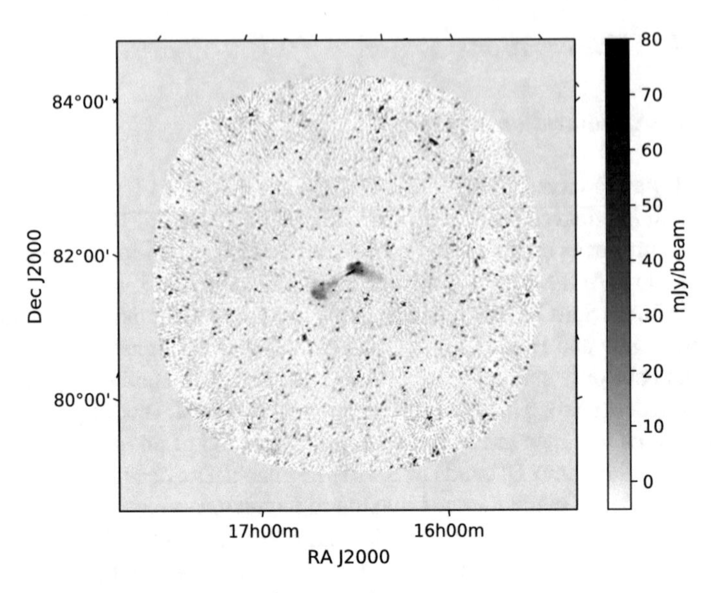

Fig. 5.1 Grayscale image showing the full primary beam corrected LOFAR HBA field of view total intensity map. The resolution is 40 arcsec and the noise is 2.0 mJy/beam

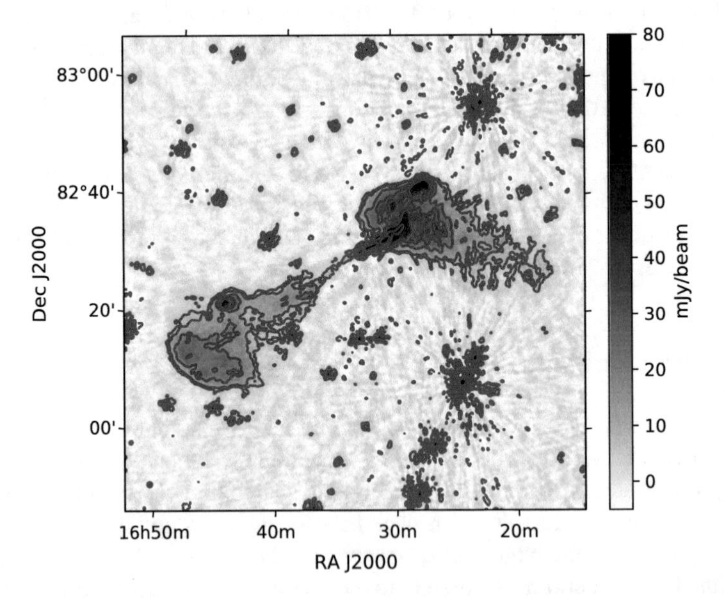

Fig. 5.2 Grayscale image showing the LOFAR HBA total intensity map. Contours are at −3, 3, 5, 10, 15, 20, 60, 150, 200, 400 × $\sigma_{\rm rms}$ where $\sigma_{\rm rms} = 2.0$ mJy/beam. The blue ellipse in the bottom left hand corner shows the 40 arcsec beam

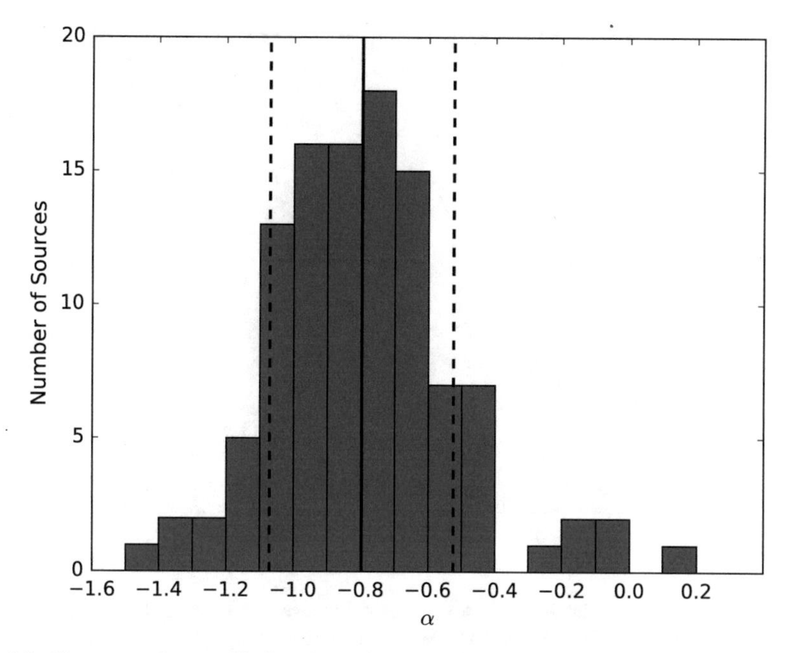

Fig. 5.3 Histogram of spectral indices for each source within the LOFAR HBA half power point of the primary beam with an NVSS counterpart for the weighted averaged HBA image. The solid black line represents the average spectral index of $\alpha = -0.8 \pm 0.3$ and the broken black lines represent the 1σ boundaries

First a catalogue of sources is generated for my HBA field using PYBDSF.[3] From this catalogue bright sources with fluxes >0.1 Jy were cross matched with the VLA Low-Frequency Sky Survey (VLSS) (Lane et al. 2012) and the NRAO VLA Sky Survey (NVSS) Condon et al. (1998b). The final catalogue of sources contained only those with sources with both a VLSS counter-part and an NVSS counterpart. A flux correction factor was then found for each band. The correction factor was then applied to the LOFAR field and a new source catalogue was generated for the field. In order to test the reliability of the flux correction, I found the spectral index for ever source within the half power point of the LOFAR primary beam with an NVSS counterpoint. Figure 5.3 shows the histogram of the spectral indices for different LOFAR frequencies. The solid black line shows the average spectral index with the broken black lines representing the 1σ boundaries.

[3]PYBDSF documentation: http://www.astron.nl/citt/pybdsm/.

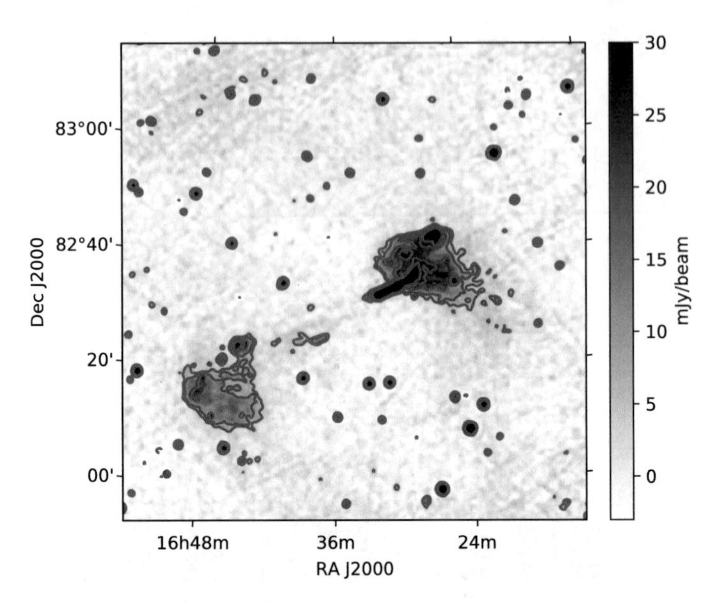

Fig. 5.4 Greyscale image shows the 325 MHz image previously published in Mack et al. (1997). The image has a resolution of 55 arcsec and an rms noise of $\sigma_{\rm rms} = 2$ mJy/beam. $-3, 3, 5, 10, 15, 20 \times \sigma_{\rm rms}$

5.2.2 Archival Data

I have used a number of archival data sets in the analysis of NGC 6251. From Mack et al. (1997) I have used the WSRT 325 MHz images and the 610 MHz images as well as the Effelsberg 10 GHz images. For more information on these data refer to Mack et al. (1997, 1998). Figures 5.4 and 5.5 show the previously published WSRT 325 MHz image and WSRT 610 MHz image, kindly provided by Karl-Heinz Mack.

A number of datasets from the NRAO archive, details of which are summarised in Table 5.2, were also used. The VLA datasets used were chosen to best match the resolution of the LOFAR observations. Observations at 8 GHz in D configuration as well as 1.4 GHz and 325 MHz in B configuration were used to image the core of NGC 6251. Observations at 1.4 GHz in D configuration were used to analyse the large scale structure of NGC 6251.

The B configuration P band data and D configuration L band data were imaged and reduced in CASA4.7. A simple calibration strategy was adopted for the D configuration L band data. The flux scale was set using Perley and Butler (2013). An initial phase calibration was performed using the flux calibrator followed by the bandpass calibration and a final amplitude and phase calibration. NGC 6251 was observed as two pointings, one centred on the core and the other on the southern lobe. Both pointings were imaged in two steps. The first round of imaging included no multiscale and a mask which excluded large extended regions. Once all compact emission or narrow emission, such as the jet, was included in the model, a second round of imaging was

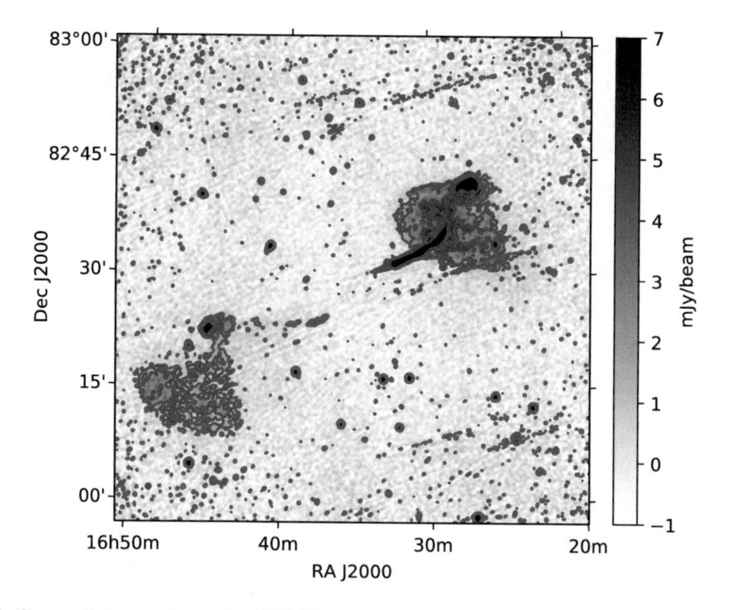

Fig. 5.5 Greyscale image shows the 610 MHz image previously published in Mack et al. (1997). The image has a resolution of 28 arcsec and an rms noise of $\sigma_{rms} = 0.4$ mJy/beam. −3, 3, 5, 10, 15, 20 × σ_{rms}

Table 5.2 Archival VLA data. Column one shows the proposal ID, column two shows the date the observation was taken, column three shows the VLA configuration, column four shows the frequency band used, column five list the references for the images used

Proposal ID.	Date	Configuration	Band	References
AK461	5-Oct-1998	B	P	This chapter
VJ49, VJ38	20-Nov-1988	A, B	L	Evans et al. (2005)
Test	5-Dec-1985	D	L	This chapter
AB3346	1-Dec-1985	D	L	This chapter
AM0322	9-May 1991	D	X	Evans et al. (2005)

done using multiscale in order to properly image the diffuse emission. The data were imaged with a uvrange of 140 − 4400 λ and a natural weighting scheme, The final image of the northern lobe has a resolution of 58 arcsec and an rms of 650 μJy/beam. The final image of the southern lobe has a resolution of 55 arcsec and an rms of 260 μJy/beam.

The VLA P band data was calibrated using a similar procedure with one additional step at the start. GPS data was used to generate a TEC map and this was used to apply a phase correction to the data using the CASA task gencal. The final image, shown

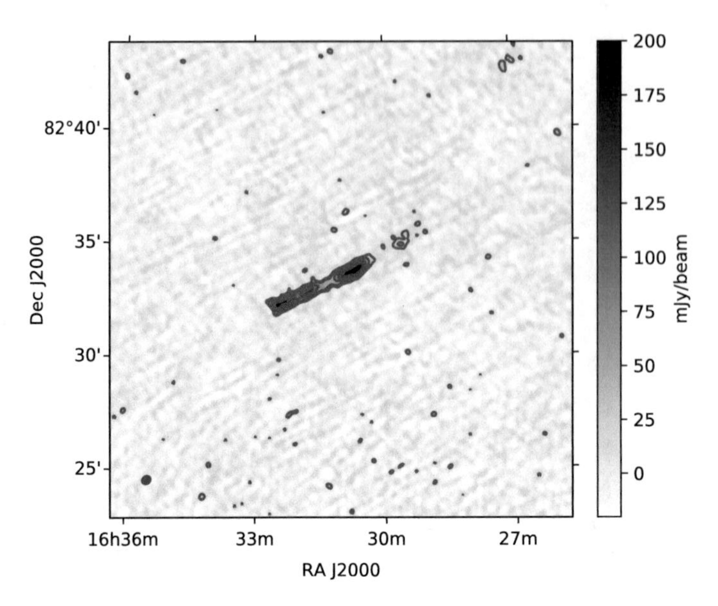

Fig. 5.6 Greyscale image of the P band VLA B configuration images of NGC 6251. VLA contours are in red at 3, 5, 10, 15, 20 $\times$ $\sigma_{\rm rms}$ where $\sigma_{\rm rms} = 8$ mJy

Table 5.3 Summary of NGC 6251 images

Array	frequency	Resolution (arcsec)	$\sigma_{\rm rms}$ (mJy/beam)
LOFAR HBA	150 MHz	40	2.0
WSRT	325 MHz	55	2.0
VLA B config.	325 MHz	20	7.0
WSRT	610 MHz	28	0.4
VLA D config. (NL)	1.4 GHz	58	0.65
VLA D config. (SL)	1.4 GHz	55	0.26
Effelsberg	10 GHz	69	1.0

in Fig. 5.6 had a resolution of 20 arcsec and an rms noise of 8 mJy/beam. Table 5.3 is a summary of the resolution and rms noise of the images used in this paper.

5.3 Results

5.3.1 Total Intensity Maps

Table 5.4 shows the flux densities measured within the 3σ contour line in various regions of NGC 6251 for different frequencies between 150 MHz and 10 GHz.

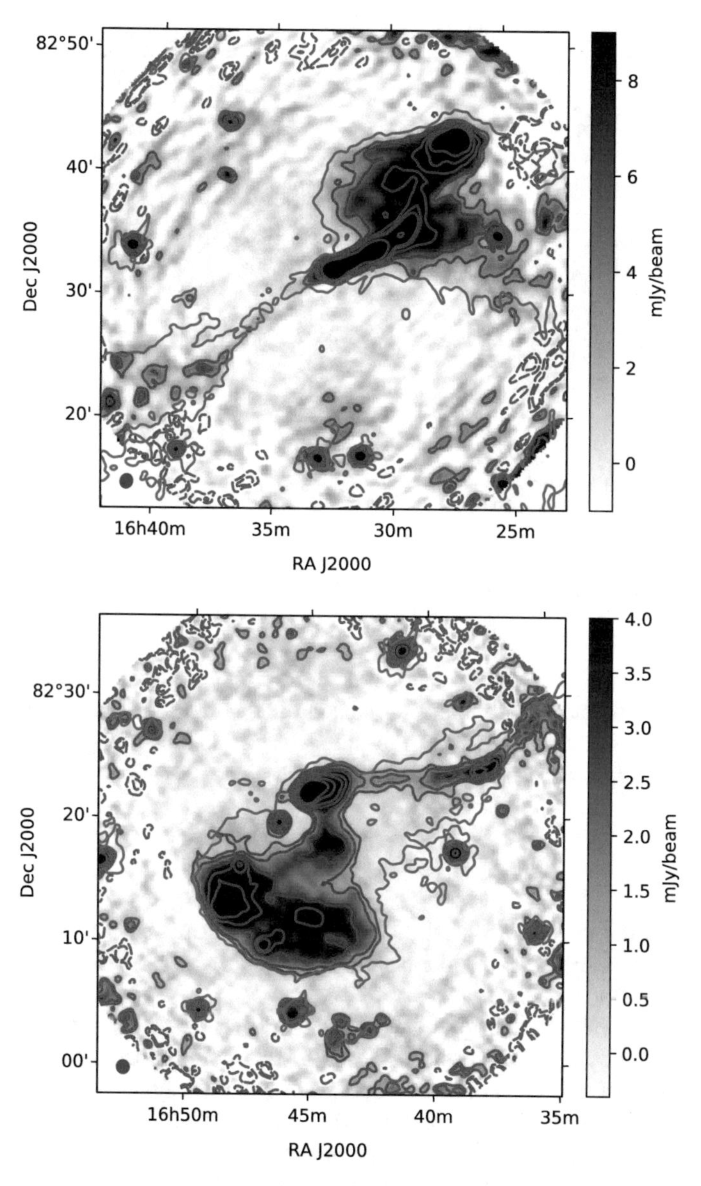

Fig. 5.7 Greyscale image of the 1.4 GHz VLA D configuration images of the northern and southern lobe of NGC 6251. VLA contours are in red at −5, −3, 5, 10, 20, 30, 40, 100 × $\sigma_{\rm rms}$ where $\sigma_{\rm rms}$ = 1.0 mJy. LOFAR HBA contours are shown in blue at $3\sigma_{\rm rms}$ where $\sigma_{\rm rms}$ = 2.0 mJy

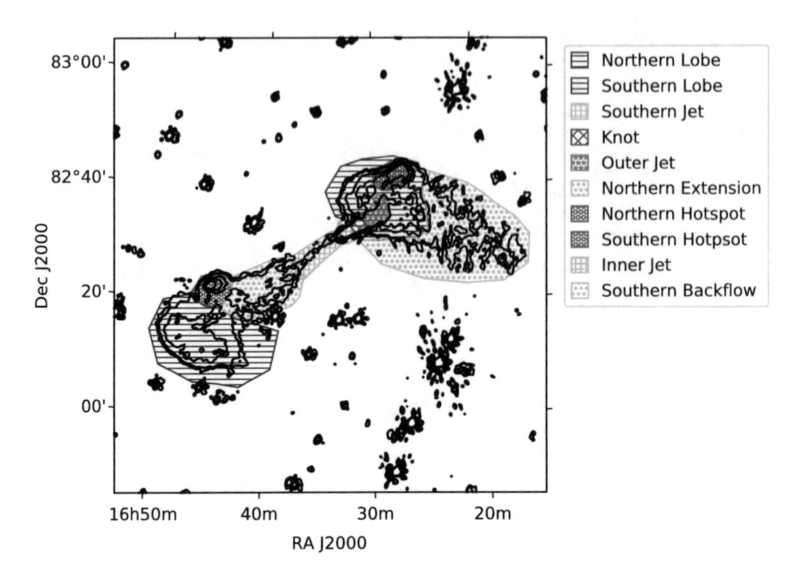

Fig. 5.8 Map of the regions used to measure the fluxes for NGC 6251

Figure 5.8 shows a map of the regions used. Point sources embedded in the lobe emission were replaced with blanked pixels. Point source subtraction was not used as I did not have access to the uvdata for the 610 and 325 MHz images. Errors in the flux measurements were calculated using the equation:

$$\sigma_{S_\nu} = \sqrt{(\sigma_{\rm cal} S_\nu)^2 + \left(\sigma_{\rm rms}\sqrt{N_{\rm beam}}\right)^2}, \tag{5.1}$$

where $\sigma_{\rm cal}$ is the uncertainty in the calibration of the flux-scale and $N_{\rm beam}$ is the number of independent beams in the source. The uncertainty in the calibration was taken to be $\sim$10%.

In the case of the southern jet a 3σ upper limit was used for the 325, 610 and 1500 MHz data. A 3σ upper limit was used for the northern extension at 325 and 610 MHz. The VLA L band data did not cover the region of the northern extension.

Figure 5.2 shows the full bandwidth HBA image of NGC 6251. This image was generated from a weighted average of the inband images.

The main jet extends north-west from the core with a bright knot at 200 arcsec (99.6 kpc) from the core. The jet then bends north and terminates at a hotspot. The northern lobe is coincident with the hotspot and jet down as far as the knot. A diffuse low surface brightness component extends west from the lobe. This extension was detected in the 325 MHz map of Mack et al. (1997). The 150 MHz LOFAR data in this chapter show that the region extends a further 14.4 arcmin or 430 kpc for a total length of 19 arcmin or 581 kpc.

The counter jet is detected at a 3σ level in the HBA image shown in Fig. 5.2, the clearest detection of the counter jet on these scales to date. The counter jet extends

Table 5.4 Flux densities measured for various regions of NGC 6251 at different frequencies. Column 1 shows the region, column 2 shows the flux density measured at 150 MHz, column 3 shows the flux density measured at 325 MHz, column 4 shows the flux density measured at 625 MHz, column 5 shows the flux density measured at 1.5 GHz and column 6 shows the flux density measured at 10 GHz

Component	$S_{150\,MHz}$ (Jy)	$S_{325\,MHz}$ (Jy)	$S_{610\,MHz}$ (Jy)	$S_{1.5\,GHz}$ (Jy)	$S_{8\,GHz}$ (Jy)	$S_{10\,GHz}$ (Jy)
Core region	0.2 ± 0.1[a]	0.27 ± 0.04[b]	0.32 ± 0.07[a]	0.44	0.72 ± 0.03	0.7 ± 0.3[a]
Inner jet (including core)	2.2 ± 0.2	1.4 ± 0.1	1.2 ± 0.1	0.9 ± 0.1	–	0.81 ± 0.08
Inner jet (core subtracted)	1.9 ± 0.2	1.2 ± 0.1	0.9 ± 0.1	0.5 ± 0.1	–	0.1 ± 0.3
Knot	2.8 ± 0.3	1.7 ± 0.2	1.2 ± 0.1	0.75 ± 0.08	–	0.22 ± 0.02
Outer jet	1.8 ± 0.2	0.90 ± 0.09	0.56 ± 0.06	0.34 ± 0.04	–	0.072 ± 0.008
Northern lobe	6 ± 1	2.0 ± 0.5	1.0 ± 0.3	0.5 ± 0.1	–	–
Northern extension	2.6 ± 0.3	<0.32	<0.28	–	–	–
Northern hotspot	2.0 ± 0.2	0.93 ± 0.09	0.59 ± 0.06	0.35 ± 0.04	–	0.083 ± 0.009
Southern jet	0.26 ± 0.03	<0.09	<0.08	–	–	–
Southern backflow	1.8 ± 0.2	<0.169	<0.1	0.05 ± 0.01	–	–
Southern lobe	5.8 ± 0.6	1.5 ± 0.2	0.51 ± 0.05	0.44 ± 0.06	–	–
Southern hotspot	1.0 ± 0.1	0.30 ± 0.03	0.20 ± 0.02	0.15 ± 0.02	–	0.022 ± 0.003

[a]Predicted using core spectral index. calculated from VLA P, L and X band data.
[b]Measured from VLA P band data

to the south east. At 713 arcsec (355 kpc) from the core, the jet bends to the east. The bend is bright and detected at 325, 610 and 1400 MHz. The VLA L band image in Fig. 5.7b shows the brightened jet continues eastward in a very linear fashion until it reaches a bright compact hotspot. The jet is again deflected at the hotspot and moves to the south east before terminating in a well confined southern lobe.

A region of diffuse low surface brightness emission can be seen coincident with the southern jet. This emission was previously only seen in the Waggett et al. (1977) 150 MHz map. As such this is very steep spectrum emission and is likely lobe emission which has been deflected back towards the core.

5.3.1.1 Spectral Index

The spectral index was calculated for each region of NGC6251 using the fluxes shown in Table 5.4. Figure 5.9 shows the best fit power law to the integrated flux density versus frequency for each region and Table 5.5 lists the fitted spectral indices.

The core of NGC6251 is not a resolved region in the HBA, WSRT or low resolution L band images. The flux from the core contributes to the inner jet region. In order to subtract the core contribution from the inner jet region, archival P, L and X band VLA data were used. Table 5.4 shows the core fluxes measured from each of these datasets. The spectral index of the core as measured form these data is $\alpha = 0.3 \pm 0.1$ Where $S_{\nu} \propto \nu^{\alpha}$. Using this spectral index the core flux was predicted for each of my datasets and subtracted from the integrated flux of the inner jet region.

In order to look at the variation of the spectral index across the source and remove any ambiguity due to inconsistent uvcoverage, the HBA data was reimaged with a uvrange of $140 - 4400\ \lambda$ matching that of the VLA. The resulting HBA image has an rms of 1.5 mJy/beam. A spectral index map was made from 150 MHz to 1500 MHz using a 7σ limit. The resulting images are shown in Fig. 5.10.

The spectral index map shows a flat core. The spectral index of the main jet is ~ -0.5 down the spine, steeping on either side to -0.8. The jet steepens as it enters the lobe to $\sim -0.6/-0.7$ before flattening to $\sim -0.4/0.5$ in the hotspot. The spectral index of the lobe varies from $-0.7/-0.8$ near the jet/hotspot to < -1 towards the extension.

The extension of the northern lobe is outside the primary beam of the VLA and the 610 MHz WSRT is likely not sensitive to emission on such large scales. The 325 MHz data should be sensitive to emission on these scales. The fact that the HBA image shows the extension continuing for another 14.4 arcmin past the WSRT 325 MHz boundary suggests that the emission has a very steep spectral index, at least steeper than $\alpha = -2.7$. This emission is likely very old.

The base of the counter jet can be seen in the spectral index map of the Northern lobe. The spectral index is between -0.6 and -0.7, steeper than in the main jet. This appears to be the flattest part of the pre-bend region. Of the counter jet. Beyond 60 kpc the counter jet has steepened such that it is only visible at 150 MHz.

The southern jet reappears at higher frequencies in what appears to be a bend. The spectral index of this bend is ~ -0.5 with a cocoon of steeper emission (< -0.9) surrounding it.

The spectral index for the hotspot is flat with $\alpha \sim -0.4/0.5$. The region of the southern lobe visible in the spectral index map is flatter than that seen for the northern lobe with $\alpha \sim -0.6$.

Similar to the extension of the northern lobe, the diffuse low surface brightness emission seen around the southern jet in the HBA image is not seen in the 325 MHz WSRT image suggesting that the spectral index is at least as steep as -1.6. This is steeper than the spectral index seen in the lobe reinforcing the idea that his is ageing material from the lobe being redirected back along the jet axis.

There are some large discrepancies between integrated spectral indices and the spectral index map of the southern region of the source. These discrepancies are due

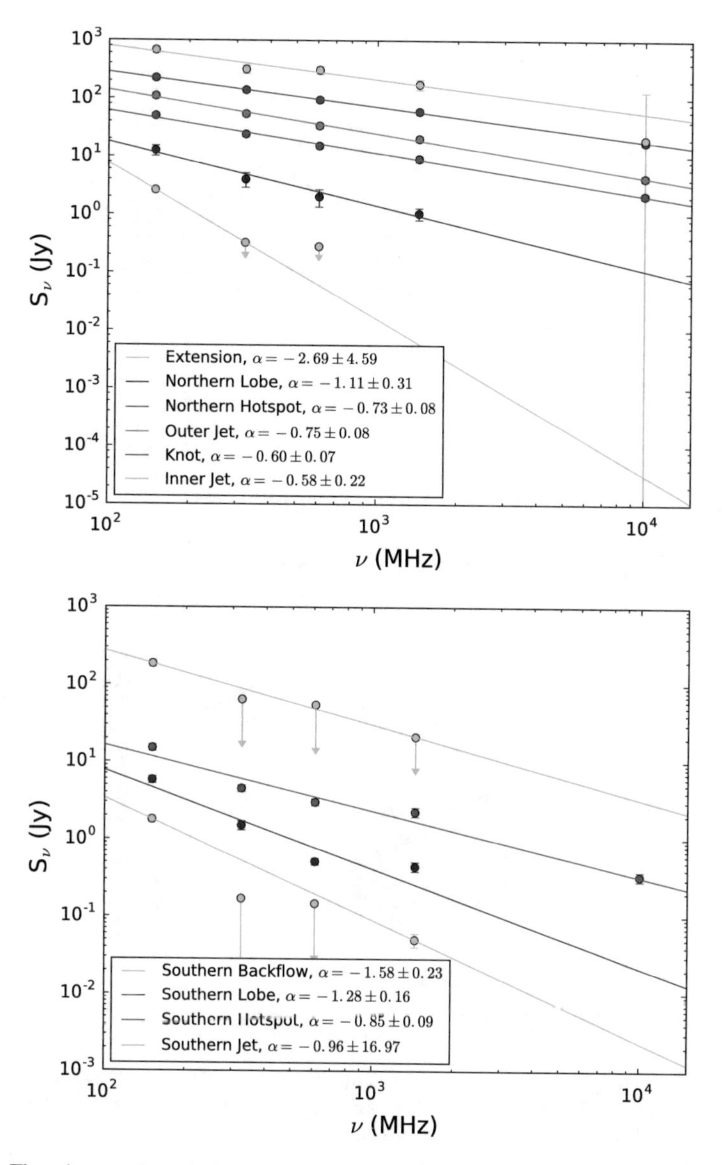

Fig. 5.9 These images show the integrated spectral index for each region in NGC6251 bar the core region. Fluxes are arbitrarily scaled to fit on the plot

to the mismatched uvcoverages between the different total intensity maps as well as overlap between the more recently accelerated emission of the hotspot and the older steeper emission of the lobe which us detected in the HBA image.

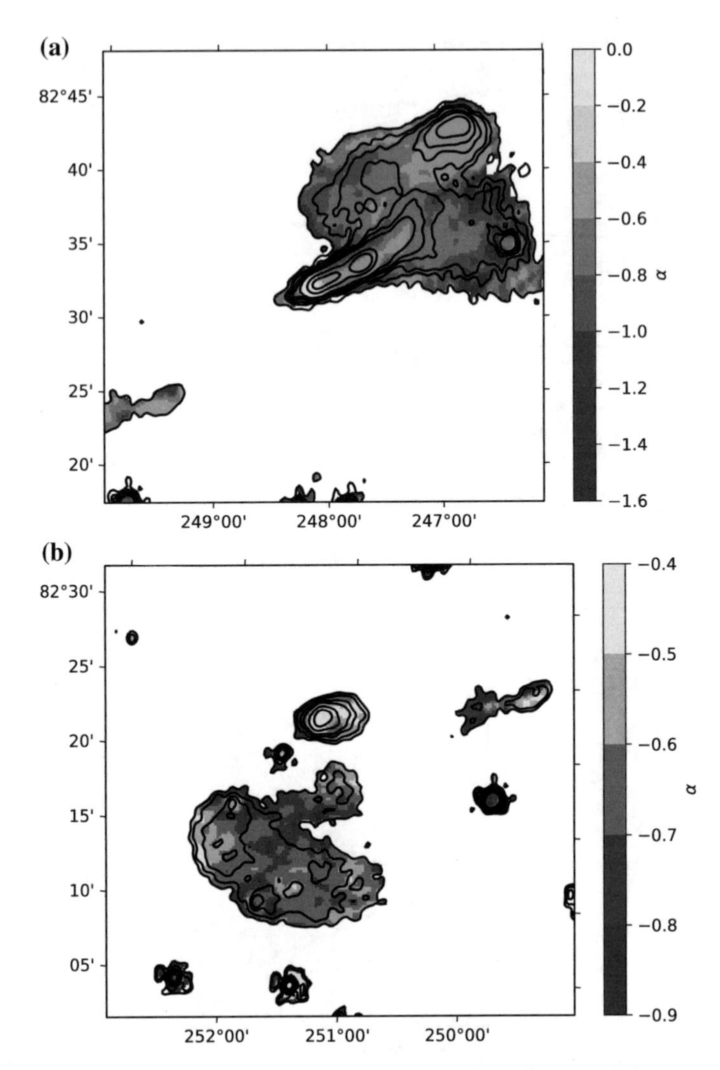

Fig. 5.10 These images show the spectral index maps from 150 MHz to 1.5 GHz for **a** the northern lobe of NGC 6251 and **b** the southern lobe of NGC 6251. The flux cuttoff used was 10.5 mJy/beam or $7\sigma_{\rm rms}$ where $\sigma_{\rm rms} = 1.5$ mJy/beam, the rms noise of the HBA image

5.3.2 *Internal Pressure*

5.3.2.1 Method

The internal pressure, $P_{\rm int}$, of a relativistic plasma is

$$P_{\rm int} = \frac{1}{3}U_{\rm e} + \frac{1}{3}U_{\rm B} + \frac{1}{3}U_{\rm p}, \tag{5.2}$$

Table 5.5 Spectral index and pressures as measured for different regions in NGC 6251. Column 1 is the region of NGC 6251, column 2 is the spectral index measured for that region, column 3 is the shape assumed when calculating the volume, column 4 is the volume of each region, column 5 is the internal pressure calculated for each region, column 5 is the external pressure measured at the projected distance of each region and column 6 is the ratio of the internal pressure and external pressure for each region

Component	α	shape	Volume (m^3)	P_{int} ($J\ m^{-3}$)	P_{ext} ($J\ m^{-3}$)	$\frac{P_{ext}}{P_{int}}$
Core region	0.3 ± 0.1	–	–	–	–	–
Inner jet (core subtracted)	-0.6 ± 0.2	Cylindrical	1.07×10^{64}	1.7×10^{-14}	4.8×10^{-13}	28
Knot	-0.60 ± 0.07	Cylindrical	4.88×10^{63}	3.3×10^{-14}	1.2×10^{-13}	4
Outer jet	-0.75 ± 0.08	Cylindrical	9.84×10^{63}	1.9×10^{-14}	2.4×10^{-14}	1.2
Northern lobe	-1.1 ± 0.3	Spherical	1.55×10^{66}	3.7×10^{-15}	7.9×10^{-15}	2
Northern extension	-2.7[a]	Cylindrical	9.1×10^{65}	4.9×10^{-15}	4.9×10^{-16}	0.1
Northern hotspot	-0.73 ± 0.08	Elliptical	5.43×10^{63}	2.7×10^{-14}	3.4×10^{-15}	0.1
Southern Jet	-1.4[a]	Cylindrical	3.24×10^{64}	3.4×10^{-15}	5.2×10^{-14}	15
Southern backflow	-1.6 ± 0.2	Cylindrical	1.29×10^{65}	1.6×10^{-15}	2.8×10^{-15}	1.7
Southern Lobe	-1.3 ± 0.2	Spherical	1.32×10^{66}	4.8×10^{-15}	4.9×10^{-16}	0.1
Southern hotspot	-0.85 ± 0.09	Elliptical	7.09×10^{63}	1.8×10^{-14}	9.5×10^{-16}	0.05

[a]Calculated using the HBA flux and WSRT upper limits

where U_e is the electron energy density, U_B is the magnetic field energy density and U_p is the proton energy density.

In order to calculate U_e and U_B, I used the SYNCH code as described in Hardcastle et al. (1998). Briefly SYNCH will calculate the equipartition magnetic field strength given the flux at a particular frequency for an electron population with power law index δ. A break in the power law can be included if necessary and minimum and maximum energies of the population must be provided along with the ratio of electron energy density to proton energy density. Once the magnetic field strength has been found the spectrum is integrated to calculate the electron and magnetic field energy density. Alternatively, if there is a detection of inverse Compton emission, the radio and X-ray data can be jointly fit to find the magnetic field energy without an assumption of equipartition.

For each region I used SYNCH to calculate the internal pressure under the assumption of equipartition between the magnetic field and non thermal particles.

5.3.2.2 Equipartition

Assuming equipartition between the magnetic field and the relativistic particles and that $U_{\mathrm{p}} = K_0 U_{\mathrm{e}}$, then

$$P_{\mathrm{int}} = \frac{3K_0 + 2}{3} U_{\mathrm{B}} \tag{5.3}$$

Using SYNCH and the spectral indices calculated in Sect. 5.3.1.1 I found $U_{\mathrm{B_{eq}}}$ for each region in NGC 6251. I assumed a lower energy cutoff of 5×10^6 eV and an upper energy of 5×10^{11} eV. I assumed an injection index of -0.6 and a break energy of 1×10^9 eV. At this break energy the spectrum steepens from the injection index to the observed spectral index. A higher break energy fails to fit the data. I assumed a spherical, elliptical or cylindrical volume for each region as seemed appropriate based on the HBA image. Table 5.5 shows the shapes and volumes used for each region. The results are tabulated in Table 5.5. As the northern extension and southern backflow are likely populated by a very old population of electrons, these data were fit using an upper energy cutoff of 1×10^{10} eV for the backflow and 2×10^9 eV for the extension.

In Fig. 5.11 I compare the calculated internal pressure with the external pressure as measured from the thermal X-ray observations in Evans et al. (2005). I find that at equipartition the inner jet is underpressured by a factor of $\sim$28, the knot is underpressured by a factor of $\sim$4 and the northern lobe is underpressured by a factor of $\sim$2. The outer jet is at pressure balance and the northern hotspot is overpressured by a factor of $\sim$10.

For the southern part of the source, the jet is underpressured by a factor of $\sim$15 while both the southern hotspot and the southern lobe are overpressured by a factor of $\sim$20 and $\sim$10 respectively. The backflow is found to be slightly underpressured by a factor of $\sim$1.7. It should be noted that the internal equipartition pressure of the western lobe reported by Evans et al. (2005) and reproduced by Croston et al. (2008) appear to be too low due to an incorrect low frequency flux measurement used by the authors.

There are a number of uncertainties in both the internal pressure and external pressure. Projection effects introduce uncertainty in both the position of the radio components in the group environment as well as the volume of the components. The northern lobe, which has a projected distance of $\sim$330 kpc, would be at pressure balance under equipartition assumptions, if its true position in the group was $\sim$440 kpc from the group centre. For a straight jet this would require an initial angle to the line of sight of 41°. However there is clear evidence of bending in the jet and so a smaller initial angle to the line of sight could still place the northern lobe at 440 kpc. The southern lobe has a projected distance from the centre of $\sim$1000 kpc. However in order for the southern lobe to be in pressure balance at equilibrium it would need to have a distance of $\sim$400 kpc from the group centre. Thus the effect of projection on the position on the southern lobe can only increase the amount by which the southern lobe is overpressured at equipartition.

Projection effects also come into play when calculating the volume of the lobes. Evans et al. (2005) argue that the axis of the northern lobe is close to the plane of the sky based on an observed discontinuity in the X-ray surface brightness. Therefore we can expect projection effects to have a minimal impact on the volume, at least for the northern lobe. No such limits on the projection of the southern lobe exist. If there is significant bending in the southern jet the visible lobe could be the cross section of a cylinder which is aligned along the line of sight. If this were the case then a spherical volume, as assumed here, would lead to an overestimate of the pressure. However assuming a cylinder with a length of 1000 kpc and volume 7.9×10^{66} m^3 only reduces the lobe pressure to 2×10^{-15} Pa, leaving the southern lobe overpressured by a factor of 4.

A very significant source of uncertainty lies in the extrapolation of the external pressure profile. The X-ray observations only detect emission out to 150 kpc, therefore comparisons between the lobe internal pressure and external pressure rely heavily on the assumption that the external pressure profile at large radii follows the 2-component β model fit at smaller radii. If instead the external pressure flattens at larger radii then the southern lobe would not be as over pressured at equipartition as it appears here. These results are discussed in Sect. 5.4.

5.3.3 *Polarisation in NGC 6251*

5.3.3.1 Applying RM Synthesis

The commissioning of polarisation observations with LOFAR is not yet complete and so it is not currently possible to calibrate the instrumental polarisation or get the absolute polarisation angle. However it is still possible to investigate the polarisation of the source using techniques such as RM synthesis (Brentjens and de Bruyn 2005) and QU fitting (Sun et al. 2015). In order to investigate the Faraday spectrum of NGC 6251 it is necessary to first account for Faraday rotation due to the ionosphere. Variations in the ionosphere during the day will lead to different degrees of Faraday rotation throughout the observations. This will cause a smearing of any signal in Faraday space (Sotomayor-Beltran et al. 2013). RMExtract[4] was used to correct for Faraday rotation due to the ionosphere. RMExtract uses maps of the Total Electron Content (TEC) and vertical Total Electron Content (vTEC) generated from GPS data to calculate the ionospheric Faraday rotation over the LOFAR stations. There are two sources for TEC and vTEC map, namely CODE and ROB. Tests during commissioning suggest that using CODE ionospheric maps recovers more of the polarised flux. As such I use the CODE maps as the input for the ionospheric correction. CODE calculates the vTEC using data from about 200 GPS/GLONASS sites of the IGS and other institutions with a time resolution of about an hour and a spatial resolution of $2.5° \times 5.0°$ (Dow et al. 2009).

---

[4]RMExtract code can be found at https://github.com/maaijke/RMextract.

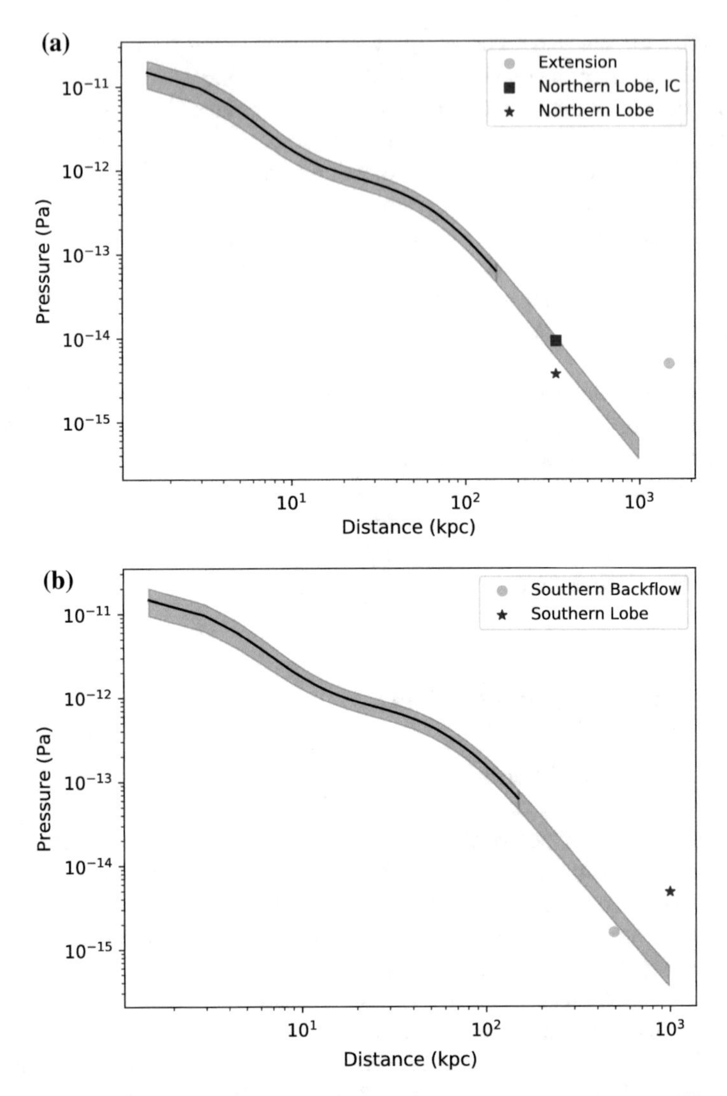

Fig. 5.11 Plot of the internal and external pressures in NGC 6251 for increasing distance from the nucleus. The solid black line shows the external pressure calculated from the thermal x-ray emission. The external pressure was previously published in Evans et al. (2005). The grey shaded region is the uncertainty in the external pressure. The hatched region represents the extrapolated external pressure where there is no direct observation of the environment. **a** The purple star shows the internal pressure of the northern lobe and the blue circle shows the internal pressure for the northern extension. The purple square shows the internal pressure of the lobe calculated from inverse compton measurements reported by Takeuchi et al. (2012). **b** The purple star shows the internal pressure of the southern lobe and the blue circle shows the internal pressure for the southern backflow

Once the RM correction has been applied to every subband and time step, individual channels were split from subbands and imaged in stokes Q and U using AWIMAGER. In order to ensure an adequate Faraday depth range, the channel width was kept at 48 kHz. An inner uvcut of 200λ, corresponding to an angular scale of $\sim$20 arcmin, was used in order to avoid imaging galactic foreground emission. The Faraday depth spectrum was calculated from the resulting Q and U images using PYRMSYNTH RMEXTRACT.[5] The maximum observable Faraday depth, $\phi_{\mathrm{max-depth}}$, the resolution in Faraday space, $\delta\phi$, and the largest scale in Faraday space that can be detected, $\phi_{\mathrm{max-scale}}$, are given by (Brentjens and de Bruyn 2005)

$$\|\phi_{\mathrm{max-depth}}\| \approx \frac{\sqrt{3}}{\delta\lambda^2} \tag{5.4a}$$

$$\delta\phi \approx \frac{2\sqrt{3}}{\Delta\lambda^2} \tag{5.4b}$$

$$\phi_{\mathrm{max-scale}} \approx \frac{\pi}{\lambda^2_{\mathrm{min}}}, \tag{5.4c}$$

where $\delta\lambda^2$ is the channel width squared, $\Delta\lambda^2$ is the width of the λ^2 distribution and λ^2_{min} is the minimum wavelength squared. For these observations this gives $\delta\phi = 0.87$ rad m^2 and $\phi_{\mathrm{max-scale}} = 0.46$ rad m^{-2}. I first search over a large, -1000 rad m^{-2}–1000 rad m^2, using a coarse Faraday depth cell size of 2 rad m^{-2}. I use this spectrum to rule out structure at high Faraday depths. I then perform Faraday rotation measure synthesis over Faraday depths of -300 rad m^{-2} to 300 rad m^{-2}. I use a cell size of 0.2 rad m^{-2} to properly sample the rotation measure transfer function (RMTF). Figure 5.12 shows the RMTF of the HBA data.

5.3.3.2 Results from RM Synthesis

Figure 5.13 shows the Faraday spectrum between $\phi = -100$ rad m^{-2} and $\phi = 100$ rad m^{-2} for both an unpolarised source and a polarised region of NGC 6251. The strong peak centred on $\sim$0 rad m^{-2} which can be seen in both Faraday spectra is the instrumental polarisation. Figure 5.14 shows the value of polarisation at the maximum in the Faraday spectrum for each pixel. The blue contours mark the regions where the peak in Faraday spectrum is about $5\sigma_{\mathrm{rms}}$, where σ_{rms} is the noise in the Faraday spectrum of that pixel. There is a clear detection of polarisation in the knot of the jet as well as some patchy structure in the northern lobe. The rest of the source is depolarised. As it is currently not possible to calibrate the instrumental polarisation I did not make a polarised intensity map. The instrumental polarisation is so dominant in these data compared with the polarised signal that a polarised intensity map would be an image of the instrumental polarisation.

Figure 5.14 also shows the Faraday spectrum for a pixel in the knot and a pixel in the lobe. The Faraday spectrum of the lobe shows a single Faraday thin component

[5]PYRMSYNTH code can be found at https://github.com/mrbell/pyrmsynth.

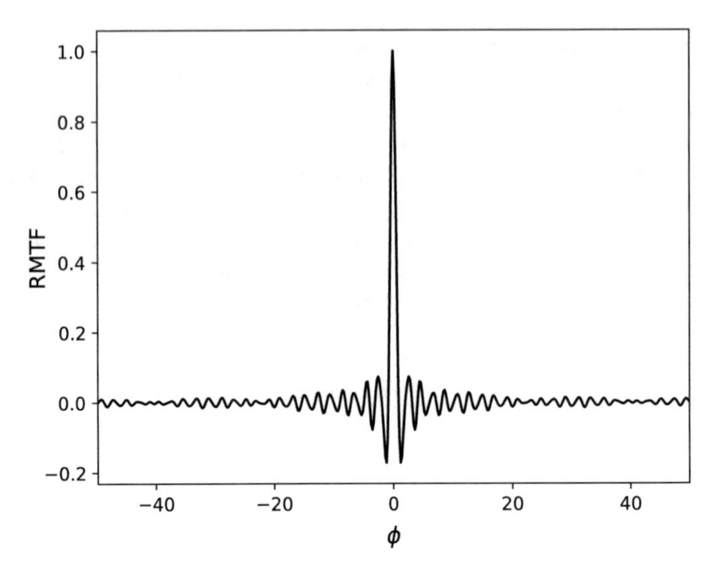

Fig. 5.12 RMTF for the LOFAR HBA data

with an average Faraday depth across the region of -54.1 rad m^{-2} with a standard deviation of 0.1 rad m^{-2}. The average amplitude of the Faraday thin component in the knot is 1.90 mJy/beam/RMTF with a standard deviation of 0.07 mJy/beam/RMTF.

Most of the knot shows a single Faraday thin component with an average Faraday depth across the region of -50.96 rad m^{-2} and a standard deviation of 0.08 rad m^{-2}. The average amplitude of this component is 3 mJy/beam/RMTF with a standard deviation of 1 mJy/beam/RMTF. Approximately 60 pixels in the centre of the knot show the double peaked structure seen in Fig. 5.14. The average Faraday depth across the region of the second peak is -52.7 rad m^{-2} with a standard deviation of 0.1 rad m^{-2}. The average amplitude of this secondary component is 1.83 mJy/beam/RMTF with a standard deviation of 0.09 mJy/beam/RMTF.

Different Faraday structure along the same line of sight can interfere in unusual ways when performing RM synthesis to produce misleading Faraday spectra Farnsworth et al. (2011). In order to further investigate the structure in the knot I used QU fitting.

Figure 5.15 is Fig. 20 as presented in Perley et al. (1984) and shows the high resolution RM maps of the northern jet of NGC 6251. Perley et al. (1984) report strong gradients in the RM of the inner jet as seen in Fig. 5.15a. These gradients lead to complete depolarisation of the signal at 150 MHz as can be seen in Fig. 5.14. Perley et al. (1984) find that beyond $\sim$180 arcsec from the core the RM is constant at -49 rad m^{-2} to within their measurement errors. This can be seen in Fig. 5.15b. This corresponds to the bright knot region in the LOFAR images and is the only strong detection of polarisation in the LOFAR HBA observations of NGC 6251. Due to

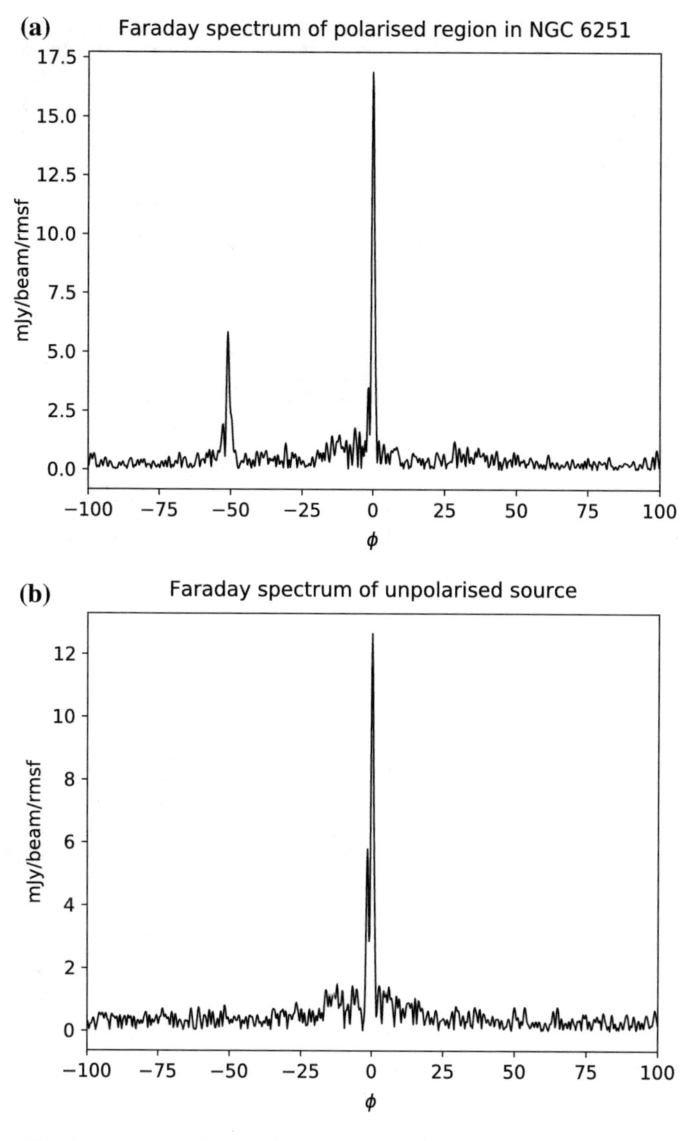

Fig. 5.13 **a** Faraday spectrum of a polarised region in NGC 6251, **b** the Faraday spectrum for an unpolarised source

LOFAR's unique resolution if Farday space it is possible to confirm that the Faraday depth is truley flat in this region with an average Faraday depth of $-50.96\,\text{rad m}^{-2}$ and a standard deviation of $0.08\,\text{rad m}^{-2}$.

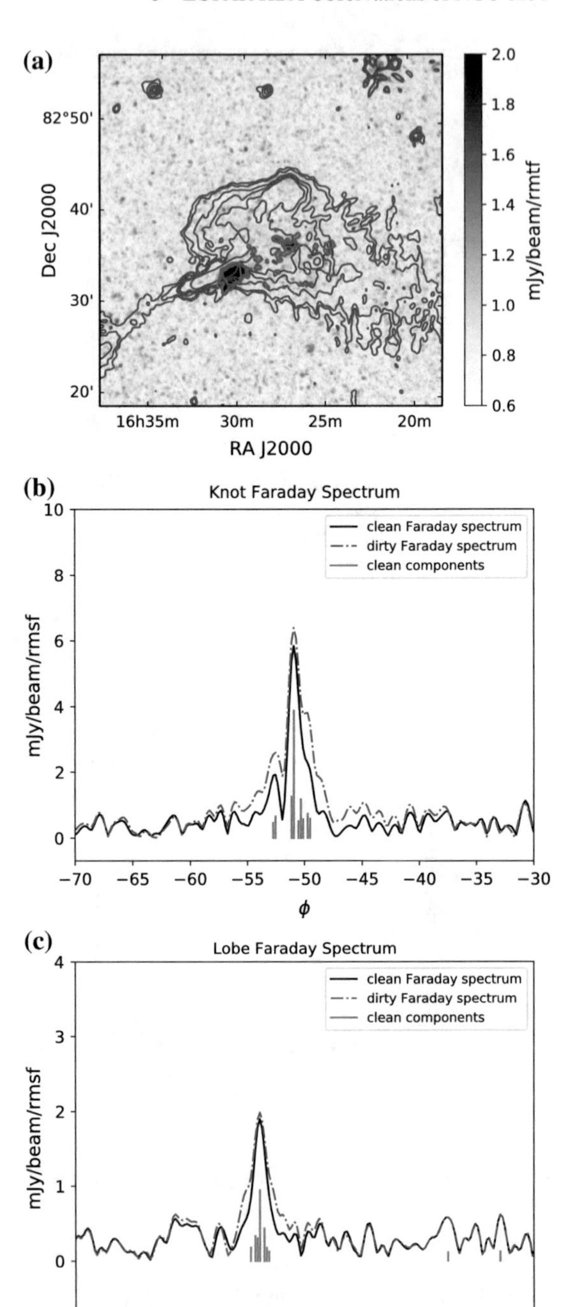

Fig. 5.14 **a** Greyscale shows the value of the peak polarisation in the Faraday depth cube. **b** The Faraday spectrum for a pixel in the knot. **c** The Faraday spectrum for a pixel in the lobe

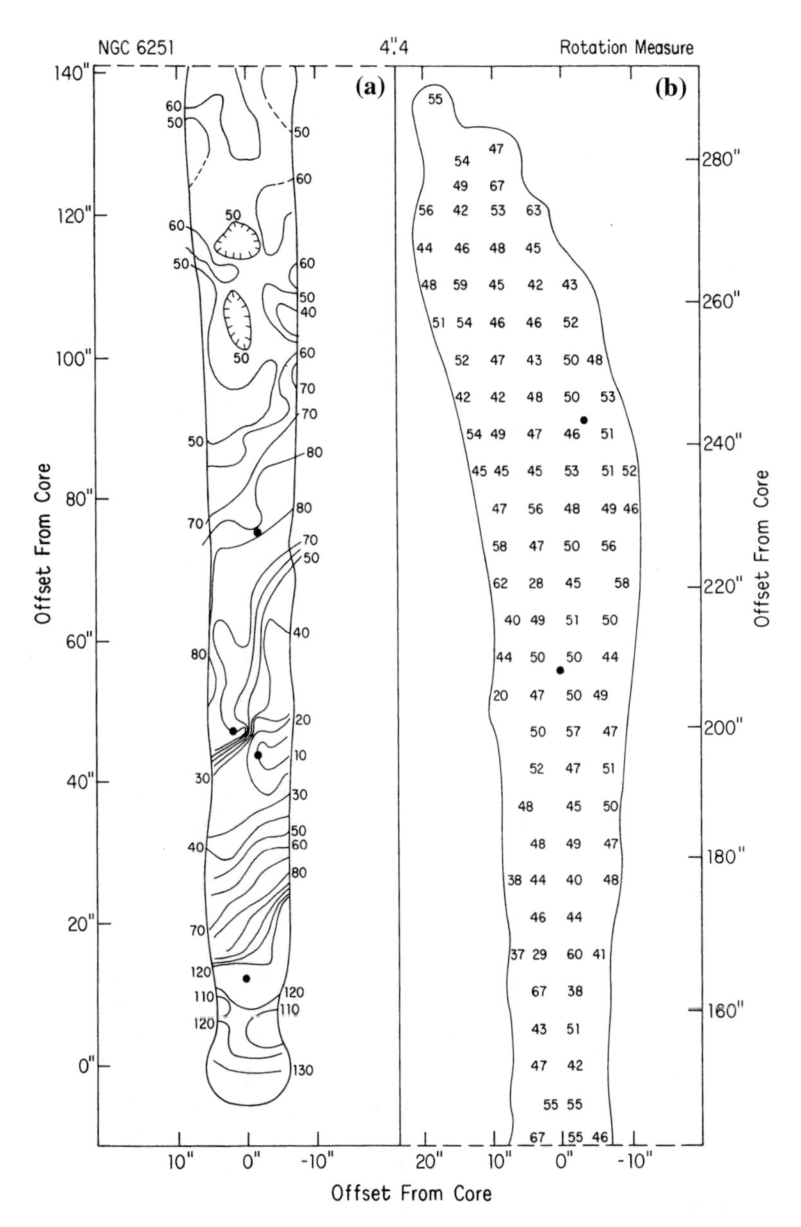

Fig. 5.15 Figure 20 reproduced from Perley et al. (1984). **a** Distribution of RM over the inner jet at 4.4 arcsec resolution. Dotted contours show regions where the RM variation is poorly determined. **b** Individual RM values in the shown at intervals of 5.25 arcsec

5.3.3.3 QU Fitting

In order to determine the exact nature and depth of components seen in the Faraday spectrum and to investigate the Faraday complexity in the source I fit models of the polarisation to the data. This process is know as QU fitting (see Sect. 2.6). For a comparison of QU fitting and RM synthesis see Sun et al. (2015). Here I use the QU- JB code presented in Sun et al. (2015). QU- JB uses a Bayesian approach to the fitting, utilising multinest to converge on solutions.

LOFAR HBA is insensitive to Faraday thick structures, resolving them out almost entirely. A Faraday thick slab observed by LOFAR would appear as 2 Faraday thin screens in the Faraday spectrum as only the edges would be visible.

I fit a number of different models to these data. First the Null hypothesis is that only instrumental polarisation is present. I fit the instrumental polarisation as a 1st order polynomial. I then fit a series of models with a 1st order polynomial and 1, 2 or 3 F thin components, hereafter, p1n1, p1n2 and p1n3 respectively. Due to LOFAR's insensitivity to Faraday thick structure I do not fit for any thick structures in the data.

In order to show the quality of the fits across the knot and the lobe we calculate the Bayes factor, $K_{\mathrm{M1}/\mathrm{M2}}$, which is given by

$$K_{\mathrm{M1}/\mathrm{M2}} = \frac{\Pr(D|M_1)}{\Pr(D|M_2)} = \frac{\int \Pr(\theta_1|M_1)\Pr(D|\theta_1, M_1)d\theta_1}{\int \Pr(\theta_2|M_2)\Pr(D|\theta_2, M_2)d\theta_2}, \tag{5.5}$$

where $\Pr(D|M_1)$ is the Bayesian evidence for Model 1, $\Pr(D|M_2)$ is the Bayesian evidence for Model 2. $\Pr(\theta_1|M_1)$ is the prior for Model 1, $\Pr(\theta_2|M_2)$ is the prior for Model 2, $\Pr(D|\theta_1, M_1)$ is the likelihood of Model 1 and $\Pr(D|\theta_2, M_2)$ is the likelihood of Model 2. Following Kass and Raftery (1995), $2\log_e K_{\mathrm{M1}/\mathrm{M2}} < 0$ is evidence for model 1, $0 < 2\log_e K_{\mathrm{M1}/\mathrm{M2}} < 2$ is only very weak evidence of model 2, $2 < 2\log_e K_{\mathrm{M1}/\mathrm{M2}} < 6$ is positive evidence for model 2, $6 < 2\log_e K_{\mathrm{M1}/\mathrm{M2}} < 10$ is strong evidence for model 2 and finally $2\log_e K_{\mathrm{M1}/\mathrm{M2}} > 10$ is very strong evidence for model 2.

Figure 5.16 shows the results for the p1n1 and p1n2 models. The evidence of the p1n3 model is less than the evidence for all other tested models for every pixel and so is not shown. Figure 5.16a shows the Faraday depth found for the p1n1 model for each pixel where $K_{\mathrm{p1n1}/\mathrm{p1n0}} > 1$. Figure 5.16c shows $2\log_e K_{\mathrm{M1}/\mathrm{M2}} > 2$, where white to blue ($2\log_e K_{\mathrm{M1}/\mathrm{M2}} > 2$) indicates support for the p1n1 model and red ($2\log_e K_{\mathrm{M1}/\mathrm{M2}} < 2$) indicates support for the Null hypothesis. I find an average Faraday depth in the knot of -50.8 rad m^{-2}. Despite the fact that there is a significant polarised region detected at 5σ in the Faraday spectra of the lobe, only 10 pixels are detected in the lobe with $K_{\mathrm{p1n1}/\mathrm{p1n0}} > 1$. The average Faraday depth of these pixels is -53.6 rad m^{-2} with a standard deviation of 0.7 rad m^{-2}, in agreement with that measured from the Faraday spectra.

Figure 5.16b and d show the Faraday depths found for the two components in the p1n2 model for each pixel where $K_{\mathrm{p1n2}/\mathrm{p1n1}} > 1$. Figure 5.16e shows $2\log_e K_{\mathrm{M1}/\mathrm{M2}}$ where white/blue indicates that the evidence for the p1n2 model is greater than the evidence for the p1n1 model. As seen in the Faraday spectra, the centre of the knot shows two Faraday thin structures. QU fitting finds that the Faraday depth of the first peak is -50.75 rad m^{-2} with a standard deviation of 0.02 rad m^{-2}, in agreement with the Faraday spectra. However QU fitting finds that the average Faraday depth of the second peak is -50.3 rad m^{-2} with a standard deviation of 0.1 rad m^{-2}. These two Faraday thin components would not produce the Faraday Spectrum shown in Fig. 5.14. I suspect that the source of this discrepancy is residual instrumental polarisation that has not been accounted for in the fitted model. This residual instrumental polarisation has a stronger presence in the data then the secondary peak and the algorithm attempts to fit the instrumental polarisation rather than the secondary peak. The prior, that the Faraday depth be between -70 and -30 rad m^{-2}, forces the algorithm to place the instrumental polarisation somewhere other than 0 rad m^{-2}. I have tested this by changing the prior so that the allowed Faraday depth is between -100 and 100 rad m^{-2}. With this prior the second Faraday thin component is found close to 0 rad m^{-2}.

Our inability to accurately model the instrumental polarisation as a simple polynomial is most likely due to the ionospheric calibration with RMEXTRACT. This has the effect of shifting the instrumental polarisation away from zero by different amounts as a function of time.

There are very small variations in Faraday depth of the order $\Delta\phi \sim 0.2$ rad m^{-2}. Figure 5.17 shows the distribution of the Faraday depths in the knot. The variance in the Faraday depth is $\sigma^2_{\rm RM} = 5 \times 10^{-3}$ rad^2 m^{-4}. If the variation in Faraday depth is due solely to noise/measurement error then the expected variance can be calculated as

$$\sigma^2_{\rm RM,noise} = \frac{\sum \sigma^2_{\#,i}}{N_{\rm pix}} \tag{5.6}$$

where σ_{ϕ_i} is the error in Faraday depth for pixel i and $N_{\rm pix}$ is the number of pixels. I find that the expected variance is $\sigma^2_{\rm RM,noise} = 2 \times 10^{-2}$ rad^2 m^{-4}. That $\sigma_{\rm RM,noise}$ is so much larger than $\sigma_{\rm RM}$ shows that the measurement errors are unrealistically larger. This is to be expected. I am unable to properly account for the instrumental polarisation in the Q and U data. This leads to a wide posterior distribution. However for the structure to be real, the estimated errors would need to be a 3 times larger than the actual uncertainty in Faraday depth.

The measured variance constitutes an upper limit on any real physical variation in Faraday depth. If it is assumed that the group environment is composed of cells with a uniform density and magnetic field strength but with a random field orientation then the expected variance in Faraday depth is

$$\sigma^2_{\rm RM} = 812^2 \Lambda_{\rm B} \int \left(B_{\parallel} n_{\rm e}\right)^2 dl. \tag{5.7}$$

where $\Lambda_{\rm B}$ is the characteristic scale length of the magnetic field and $B_{\parallel} = \frac{B}{\sqrt{3}}$ is the component of the magnetic field along the line of sight (Lawler and Dennison 1982). In order to determine a reasonable $\lambda_{\rm B}$ I make use of the results in Cho and Ryu (2009). Cho and Ryu (2009) suggest that $\Lambda_{\rm B}$ is related to the energy injection

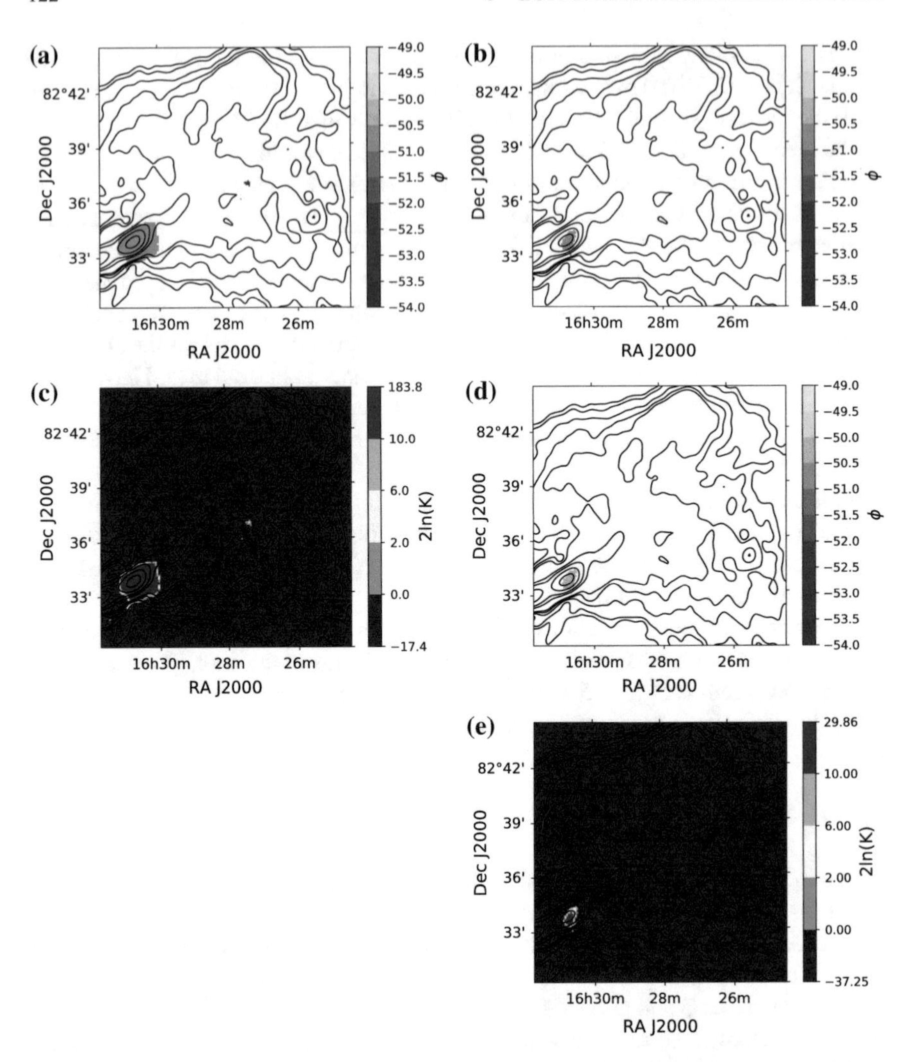

Fig. 5.16 Results of the QU fitting. **a** Shows the Faraday depth of the component found for a single Faraday thin screen plus instrumental polarisation. **b** Shows the Faraday depth of the first component found when fitting two Faraday thin screens plus instrumental polarisation. **c** The Bayes factor when comparing the fits for a single thin screen plus instrumental polarisation with just instrumental polarisation. Red indicates support for the null hypothesis while blue indicates support for the single Faraday thin screen. White indicates inconclusive. **d** The Faraday depth of the second component found when fitting the two Faraday thin screens plus instrumental polarisation. **e** The Bayes factor when comparing the single Faraday thin screen plus instrumental polarisation with the model of two Faraday thin screens plus instrumental polarisation. Red indicates support for a single screen while blue indicates support for a two screens. White indicates no real preference either way

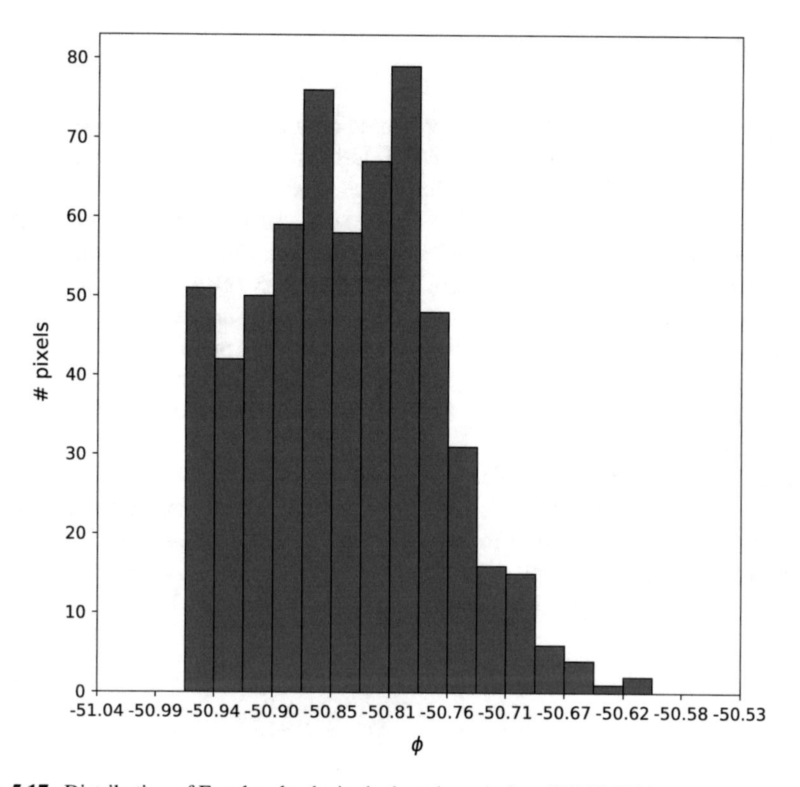

Fig. 5.17 Distribution of Faraday depths in the knot in main jet of NGC 6251

scale, L_0 by

$$\Lambda_{\mathrm{B}} = \frac{0.3}{45} L_0 \frac{t}{t_{\mathrm{eddy}}} \tag{5.8}$$

where t is the age of the Universe and t_{eddy} is the local eddy turnover time. Here we take $L_0 = r_{\mathrm{c}}$ where $r_{\mathrm{c}} = 74$ kpc is the core radius of the group environment of NGC 6251. I assume a value for $\frac{t}{t_{\mathrm{eddy}}}$ which is between that for clusters and filaments so that $\frac{t}{t_{\mathrm{eddy}}} = 20$. This gives a characteristic scale length of $\Lambda = 10$ kpc.

In order to evaluate the integral in Eq. 5.7 I must know the position of the knot in the group environment. In order to get a conservative upper limit on B I will assume an angle to the line of sight of 55°. This would place the knot at 200 kpc from the core. Under these assumptions we find that $B < 0.2\,\mu$G. These results are discussed in Sect. 5.4.

5.4 Discussion

I compare the HBA and VLA maps presented here to the 325–610 MHz map and 408–10 GHz maps in Mack et al. (1998). I find that my spectral index map agrees well with the 325–610 MHz map except for the bend in the southern jet. Here I find a spectral index of ~ -0.5 for the bend in the southern jet while Mack et al. (1998) find a spectral index < -1. The HBA image shown in Fig. 5.2 shows the backflow of the lobe material from the southern lobe. In the 325 MHz WSRT image only a small region of this structure is detected, not enough to distinguish from the jet. In the 610 MHz image only the bend is visible. I suggest that the presence of the older lobe emission, coincident with the jet has led to the steep spectral index in the Mack et al. (1998) spectral index map and that the bend is indeed a real feature of the counter jet.

The oldest regions of NGC 6251 are the extension of the Northern lobe and the backflow of the southern lobe. These regions are only clearly detected at 150 MHz. As such it is not possible to fit for a break frequency, ν_b. Instead I assume that the break frequency lies somewhere between 150 and 325 MHz. I calculate the age, t, of both regions following Alexander and Leahy (1987) so that

$$t^2 = \frac{2.52 \times 10^5 U_{\mathrm{eq}} \mu_0 2}{\nu_b \left[(2\mu_0 U_{\mathrm{eq}})^2 + B_{\mathrm{m}}^2 \right]^2}, \tag{5.9}$$

where $B_{\mathrm{m}} = 0.318\,(1+z)^2$ is the equivalent field strength of the microwave background radiation assuming the present day temperature of 2.726 K. This yields an age of 205 Myr $< t <$ 368 Myr for the northern extension and 209 Myr $< t <$ 307 Myr for the southern backflow. These regions are significantly older than the ages for both the northern lobe (<42 Myr) and the southern lobe (41 Myr) (Mack et al. 1998) and are possibly relic emission from an earlier epoch of activity.

In Sect. 5.3.2.2 the internal pressure in the lobes of NGC 6251 was calculated and compared to the external pressure. While the northern lobe was found to be marginally underpressured at equipartition, the southern lobe was found to be significantly overpressured at equipartition, requiring the assumed external pressure profile to flatten at large radii in order to achieve pressure balance in the southern lobe. It is however entirely possible that the southern lobe is over-pressured with respect to the environment. We have assumed that the group environment is symmetric about NGC 6251, however an asymmetric atmosphere could be responsible for having two lobes whose dynamics/relation to the external pressure appear to be different. It has been suggested that the asymmetries seen in radio galaxies is due to environmental effects (Pirya et al. 2012; Schoenmakers et al. 2000; Lara et al. 2004). NGC 6251 has a separation ratio of 1 : 2.2 and Chen et al. (2011) show that the galaxy overdensity is larger in the direction of the shorter main jet of NGC 6251. There is therefore reason to believe the environment of NGC 6251 could be asymmetric however the current available X-ray data is not sufficient to investigate this directly.

A separate estimate of the particle energetic comes from observations of inverse Compton emission. Takeuchi et al. (2012) fit a model to radio, X-ray and gamma ray data. They find the combined data are best fit with a magnetic field of $B = 0.37$ μG and an injection spectral index of $\alpha = -0.5$ which breaks to $\alpha = -0.75$ at $E_{\mathrm{b}} = 1.5 \times 10^{9}$ eV. This gives an energy ratio of $\frac{U_{\mathrm{e}}}{U_{\mathrm{B}}} = 45$ and an internal pressure of 8.5×10^{-15} Pa. This would place the lobe in pressure balance with the external pressure at the projected distance of the lobe from the cluster centre. The region used to calculate this pressure includes both the lobe and the hotspot. Given that the electron population in the hotspot and the lobe would be expected to have different characteristics, the pressure calculated by Takeuchi et al. (2012) is likely an overestimate and the true internal pressure is somewhere between the equipartition pressure calculated in this paper and that calculated in Takeuchi et al. (2012).

The northern extension and southern backflow are the most relaxed regions of NGC 6251 As such while it might be possible that the lobes are overpressured, these regions are more likely to be in equilibrium with the environment. The host galaxy group has an r_{200} of 875 kpc (Croston et al. 2008). The northern lobe reaches a projected distance of $\sim r_{500}$ while the southern lobe reaches beyond r_{200}. The extended structure of NGC 6251 is therefore probing the outskirts of the group environment and the LLS beyond. The internal pressure of the extension implies an environment pressure of 4.9×10^{-15} Pa and the internal pressure of the southern backflow implies an environmental pressure of 1.6×10^{-15} Pa. Malarecki et al. (2015) find similar pressures for 12 giant radio galaxies. They show that this pressure corresponds to the densest 6% of the WHIM.

I find polarised emission in the region of the bright knot in the main jet as well as a small region of patchy polarisation in the northern lobe. All polarisation in the inner part of the jet is depolarised due to RM gradients in this region of the source (Perley et al. 1984). It is likely that the majority of the Faraday rotation observed is due to our Galaxy. Estimating the exact value of the contribution is difficult. Higher frequency, high resolution data presented in Perley et al. (1984) shows that beyond 180 arcsec (89 kpc) from the core the average Faraday depth is -48.9 ± 0.2 rad m^{-2}. They suggest that this is the galactic contribution to the Faraday depth. Oppermann et al. (2015) reconstruct a map of the Galactic Faraday contribution using observations of extragalactic sources. This reconstructed map has an average Faraday rotation of -31.6 rad m^{-2} in the region of NGC 6251. In the data presented here I found 1 polarised source near NGC 6251 with a Faraday depth of ≈ -44 rad m^{-2} which is between the assumed galactic contributions from Perley et al. (1984) and Oppermann et al. (2015). These values suggest that the extragalactic contribution is of order $1 - 10$ rad m^{-2}.

The Faraday depth values measured in the knot are in good agreement with those found in Perley et al. (1984). However with the broadband capabilities of LOFAR I have detected complexity in the Faraday spectrum of the knot. Based on the arguments presented in Sect. 5.3.3.3 I will rely on the results from RM synthesis. There are a number of possible scenarios that could explain the Faraday structure seen in the knot. If the knot is a Faraday thick slab one might expect to see a double peaked structure in Faraday space. Due to LOFAR's insensitivity to Faraday thick structures, only the sharp edges of the slab would be visible in the Faraday spectrum. However you would expect the peaks to have similar amplitudes. I find that the dominant peak is 1.6 times stronger than the secondary peak.

One might also imagine that the knot could be embedded in a Faraday thick structure. The lobe could potentially act as such a Faraday thick screen as it is coincident with the knot. However again I would expect another peak for the region of the lobe in front of the knot which I do not see. Based on the pressure arguments above, the lobe does not require a significant thermal population to be in pressure balance with the environment and so it is unlikely that the lobe is Faraday thick.

I suggest that the most likely scenario is that the lobe is Faraday thin and located somewhere behind the knot. The strong peak in the Faraday spectrum would then due to the knot and the fainter peak due to the lobe located at a more negative Faraday depth. The argument is strengthened by the fact that the polarisation detected in in the lobe has a similar amplitude to the secondary peak in the knot with the peak in the knot.

In Sect. 5.3.3.3 I put an upper limit on the magnetic field in the group of $B < 0.2$ μG. The GRG NGC 315 is in a similar sparse environment to NGC 6251. Laing et al. (2006) find residual fluctuations in the Faraday depth of NGC 315 of order $1-2$ rad m^2 and suggest that for plausible assumptions for the central density and characteristic magnetic field length, the central magnetic field would be $B_0 = 0.15$ μG. This is comparable to the upper limit calculated here for NGC 6251. Denser group environments such as those of 3C449 and 3C31 have central magnetic field strengths of order a few μG (Laing et al. 2008; Guidetti et al. 2010).

5.5 Conclusions

In this chapter I have presented observations of NGC 6251 at 150 MHz with LOFAR HBA. The images presented here are the highest sensitivity and resolution images of NGC 6251 at these frequencies to date. Analysis of the low frequencies spectral index did not reveal any change in the low frequency spectra when compared with the higher frequency spectral index. NGC 6251 is found to be either at equilibrium or slightly electron dominated, similar to FRII sources. I calculated the ages of the low surface brightness extension of the northern lobe and the backflow of the southern lobe, which are only clearly visible at these low frequencies, to be 205 Myr $< t <$ 368 Myr and 209 Myr$< t <$307 Myr respectively. This could indicate that these components are relics of an earlier epoch of activity.

I presented the first detection of polarisation at 150 MHz in NGC 6251, including a weak detection of polarisation in the diffuse emission of the northern lobe. Taking advantage of the high Faraday resolution of LOFAR, I detect Faraday complexity in the knot of NGC 6251 and interpret the weaker component as emission from the lobe located behind the knot. I place an upper limit on the variance in the Faraday depth in the knot of NGC 6251 of $\sigma_{RM} < 5 \times 10^{-3}$ rad^2 m^{-4} and an upper limit on the magnetic field in the group of $B < 0.2\ \mu$G.

References

Abdo AA, Ackermann M, Ajello M, Allafort A, Antolini E, Atwood WB, Axelsson M, Baldini L, Ballet J, Barbiellini G et al (2010) Fermi large area telescope first source catalog. ApJS 188:405–436. https://doi.org/10.1088/0067-0049/188/2/405

Alexander P, Leahy JP (1987) Ageing and speeds in a representative sample of 21 classical double radio sources. MNRAS 225:1–26. https://doi.org/10.1093/mnras/225.1.1

Bhatnagar S, Cornwell TJ, Golap K, Uson JM (2008) Correcting direction-dependent gains in the deconvolution of radio interferometric images. A&A 487:419–429. https://doi.org/10.1051/0004-6361:20079284

Brentjens MA, de Bruyn AG (2005) 441:1217–1228. https://doi.org/10.1051/0004-6361:20052990

Chen R, Peng B, Strom RG, Wei J (2011) Group galaxies around giant radio galaxy NGC 6251. MNRAS 412:2433–2444. https://doi.org/10.1111/j.1365-2966.2010.18064.x

Cho J, Ryu D (2009) Characteristic lengths of magnetic field in magnetohydrodynamic turbulence. 705:L90–L94. https://doi.org/10.1088/0004-637X/705/1/L90

Condon JJ, Cotton WD, Greisen EW, Yin QF, Perley RA, Taylor GB, Broderick JJ (1998b) The NRAO VLA sky survey. AJ 115:1693–1716. https://doi.org/10.1086/300337

Croston JH, Hardcastle MJ, Birkinshaw M, Worrall DM, Laing RA (2008) An XMM-Newton study of the environments, particle content and impact of low-power radio galaxies. MNRAS 386:1709–1728. https://doi.org/10.1111/j.1365-2966.2008.13162.x

Dow JM, Neilan RE, Rizos C (2009) The international GNSS service in a changing landscape of global navigation satellite systems. J Geodesy 83:191–198. https://doi.org/10.1007/s00190-008-0300-3

Evans DA, Hardcastle MJ, Croston JH, Worrall DM, Birkinshaw M (2005) Chandra and XMM-Newton observations of NGC 6251. MNRAS 359:363–382. https://doi.org/10.1111/j.1365-2966.2005.08900.x

Fanaroff BL, Riley JM (1974) The morphology of extragalactic radio sources of high and low luminosity. MNRAS 167:31P–36P. https://doi.org/10.1093/mnras/167.1.31P

Farnsworth D, Rudnick L, Brown S (2011) Integrated polarization of sources at $\lambda \sim 1$ m and new rotation measure ambiguities. AJ 141:191. https://doi.org/10.1088/0004-6256/141/6/191

Guidetti D, Laing RA, Murgia M, Govoni F, Gregorini L, Parma P (2010) Structure of the magnetoionic medium around the Fanaroff-Riley class I radio galaxy 3C 449. A&A 514:A50. https://doi.org/10.1051/0004-6361/200913872

Hardcastle MJ, Gürkan G, van Weeren RJ, Williams WL, Best PN, de Gasperin F, Rafferty DA, Read SC, Sabater J, Shimwell TW, Smith DJB, Tasse C, Bourne N, Brienza M, Brüggen M, Brunetti G, Chyy KT, Conway J, Dunne L, Eales SA, Maddox SJ, Jarvis MJ, Mahony EK, Morganti R, Prandoni I, Röttgering HJA, Valiante E, White GJ (2016) LOFAR/H-ATLAS: A deep low-frequency survey of the Herschel-ATLAS North Galactic Pole field. ArXiv e-prints

Hardcastle MJ, Birkinshaw M, Worrall DM (1998) Magnetic field strengths in the hotspots of 3C 33 and 111. MNRAS 294:615. https://doi.org/10.1046/j.1365-8711.1998.01159.x

Heald GH, Pizzo RF, Orrú E, Breton RP, Carbone D, Ferrari C, Hardcastle MJ, Jurusik W, Macario G, Mulcahy D, Rafferty D, Asgekar A, Brentjens M, Fallows RA, Frieswijk W, Toribio MC, Adebahr B, Arts M, Bell MR, Bonafede A, Bray J, Broderick J, Cantwell T, Carroll P, Cendes Y, Clarke AO, Croston J, Daiboo S, de Gasperin F, Gregson J, Harwood J, Hassall T, Heesen V, Horneffer A, van der Horst AJ, Iacobelli M, Jelić V, Jones D, Kant D, Kokotanekov G, Martin P, McKean JP, Morabito LK, Nikiel-Wroczyński B, Offringa A, Pandey VN, Pandey-Pommier M,

Pietka M, Pratley L, Riseley C, Rowlinson A, Sabater J, Scaife AMM, Scheers LHA, Sendlinger K, Shulevski A, Sipior M, Sobey C, Stewart AJ, Stroe A, Swinbank J, Tasse C, Trüstedt J, Varenius E, van Velzen S, Vilchez N, van Weeren RJ, Wijnholds S, Williams WL, de Bruyn AG, Nijboer R, Wise M, Alexov A, Anderson J, Avruch IM, Beck R, Bell ME, van Bemmel I, Bentum MJ, Bernardi G, Best P, Breitling F, Brouw WN, Brüggen M, Butcher HR, Ciardi B, Conway JE, de Geus E, de Jong A, de Vos M, Deller A, Dettmar RJ, Duscha S, Eislöffel J, Engels D, Falcke H, Fender R, Garrett MA, Grießmeier J, Gunst AW, Hamaker JP, Hessels JWT, Hoeft M, Hörandel J, Holties HA, Intema H, Jackson NJ, Jütte E, Karastergiou A, Klijn WFA, Kondratiev VI, Koopmans LVE, Kuniyoshi M, Kuper G, Law C, van Leeuwen J, Loose M, Maat P, Markoff S, McFadden R, McKay-Bukowski D, Mevius M, Miller-Jones JCA, Morganti R, Munk H, Nelles A, Noordam JE, Norden MJ, Paas H, Polatidis AG, Reich W, Renting A, Röttgering H, Schoenmakers A, Schwarz D, Sluman J, Smirnov O, Stappers BW, Steinmetz M, Tagger M, Tang Y, ter Veen S, Thoudam S, Vermeulen R, Vocks C, Vogt C, Wijers RAMJ, Wucknitz O, Yatawatta S, Zarka P (2015) The LOFAR Multifrequency Snapshot Sky Survey (MSSS). I. Survey description and first results. A&A 582:A123. https://doi.org/10.1051/0004-6361/201425210

Kass RE, Raftery AE (1995) Bayes factors. J Am Stat Assoc 90(430):773–795. https://doi.org/10.1080/01621459.1995.10476572

Laing RA, Canvin JR, Cotton WD, Bridle AH (2006) Multifrequency observations of the jets in the radio galaxy NGC315. MNRAS 368:48–64. https://doi.org/10.1111/j.1365-2966.2006.10099.x

Laing RA, Bridle AH, Parma P, Feretti L, Giovannini G, Murgia M, Perley RA (2008) Multifrequency VLA observations of the FR I radio galaxy 3C 31: morphology, spectrum and magnetic field. MNRAS 386:657–672. https://doi.org/10.1111/j.1365-2966.2008.13091.x

Lane WM, Cotton WD, Helmboldt JF, Kassim NE (2012) VLSS redux: software improvements applied to the very large array low-frequency sky survey. Radio Sci 47. https://doi.org/10.1029/2011RS004941. (RS0K04)

Lara L, Giovannini G, Cotton WD, Feretti L, Marcaide JM, Márquez I, Venturi T (2004) A new sample of large angular size radio galaxies. III. Statistics and evolution of the grown population. A&A 421:899–911. https://doi.org/10.1051/0004-6361:20035676

Lawler JM, Dennison B (1982) On intracluster Faraday rotation II—statistical analysis. ApJ 252:81–91. https://doi.org/10.1086/159536

Mack KH, Klein U, O'Dea CP, Willis AG (1997) Multi-frequency radio continuum mapping of giant radio galaxies. A&AS 123. https://doi.org/10.1051/aas:1997166

Mack KH, Klein U, O'Dea CP, Willis AG, Saripalli L (1998) Spectral indices, particle ages, and the ambient medium of giant radio galaxies. A&A 329:431–442

Malarecki JM, Jones DH, Saripalli L, Staveley-Smith L, Subrahmanyan R (2015) Giant radio galaxies–II. Tracers of large-scale structure. MNRAS 449:955–986. https://doi.org/10.1093/mnras/stv273

Nolan PL, Abdo AA, Ackermann M, Ajello M, Allafort A, Antolini E, Atwood WB, Axelsson M, Baldini L, Ballet J et al (2012) Fermi large area telescope second source catalog. ApJS 199:31. https://doi.org/10.1088/0067-0049/199/2/31

Offringa AR, van de Gronde JJ, Roerdink JBTM (2012) A morphological algorithm for improved radio-frequency interference detection. A&A 539

Oppermann N, Junklewitz H, Greiner M, Enßlin TA, Akahori T, Carretti E, Gaensler BM, Goobar A, Harvey-Smith L, Johnston-Hollitt M, Pratley L, Schnitzeler DHFM, Stil JM, Vacca V (2015) 575:A118. https://doi.org/10.1051/0004-6361/201423995

Perley RA, Butler BJ (2013) An accurate flux density scale from 1 to 50 GHz. ApJS 204:19. https://doi.org/10.1088/0067-0049/204/2/19

Perley RA, Bridle AH, Willis AG (1984) High-resolution VLA observations of the radio jet in NGC 6251. ApJS 54:291–334. https://doi.org/10.1086/190931

Pirya A, Saikia DJ, Singh M, Chandola HC (2012) A study of the environments of large radio galaxies using SDSS. MNRAS 426:758–763. https://doi.org/10.1111/j.1365-2966.2012.21656.x

Saunders R, Baldwin JE, Pooley GG, Warner PJ (1981) The radio jet in NGC 6251. MNRAS 197:287–300. https://doi.org/10.1093/mnras/197.2.287

Scaife AMM, Heald GH (2012) A broad-band flux scale for low-frequency radio telescopes. MNRAS 423:L30–L34. https://doi.org/10.1111/j.1745-3933.2012.01251.x

Schoenmakers AP, Mack KH, de Bruyn AG, Röttgering HJA, Klein U, van der Laan H (2000) A new sample of giant radio galaxies from the WENSS survey–II. A multi-frequency radio study of a complete sample: properties of the radio lobes and their environment. A&As 146:293–322. https://doi.org/10.1051/aas:2000267

Sotomayor-Beltran C, Sobey C, Hessels JWT, de Bruyn G, Noutsos A, Alexov A, Anderson J, Asgekar A, Avruch IM, Beck R, Bell ME, Bell MR, Bentum MJ, Bernardi G, Best P, Birzan L, Bonafede A, Breitling F, Broderick J, Brouw WN, Brüggen M, Ciardi B, de Gasperin F, Dettmar RJ, van Duin A, Duscha S, Eislöffel J, Falcke H, Fallows RA, Fender R, Ferrari C, Frieswijk W, Garrett MA, Grießmeier J, Grit T, Gunst AW, Hassall TE, Heald G, Hoeft M, Horneffer A, Iacobelli M, Juette E, Karastergiou A, Keane E, Kohler J, Kramer M, Kondratiev VI, Koopmans LVE, Kuniyoshi M, Kuper G, van Leeuwen J, Maat P, Macario G, Markoff S, McKean JP, Mulcahy DD, Munk H, Orru E, Paas H, Pandey-Pommier M, Pilia M, Pizzo R, Polatidis AG, Reich W, Röttgering H, Serylak M, Sluman J, Stappers BW, Tagger M, Tang Y, Tasse C, ter Veen S, Vermeulen R, van Weeren RJ, Wijers RAMJ, Wijnholds SJ, Wise MW, Wucknitz O, Yatawatta S, Zarka P (2013) Calibrating high-precision Faraday rotation measurements for LOFAR and the next generation of low-frequency radio telescopes. A&A 552:A58. https://doi.org/10.1051/0004-6361/201220728

Stoffel H, Wielebinski R (1978) Observations of the very large galaxies NGC 315 and NGC 6251 at 11.1 CM. A&A 68:307–309

Sun XH, Rudnick L, Akahori T, Anderson CS, Bell MR, Bray JD, Farnes JS, Ideguchi S, Kumazaki K, O'Brien T, O'Sullivan SP, Scaife AMM, Stepanov R, Stil J, Takahashi K, van Weeren RJ, Wolleben M (2015) Comparison of algorithms for determination of rotation measure and faraday structure. I. 1100–1400 MHz. AJ 149:60. https://doi.org/10.1088/0004-6256/149/2/60

Takeuchi Y, Kataoka J, Stawarz Ł, Takahashi Y, Maeda K, Nakamori T, Cheung CC, Celotti A, Tanaka Y, Takahashi T (2012) Suzaku X-ray imaging of the extended lobe in the giant radio galaxy NGC 6251 associated with the Fermi-LAT source 2FGL J1629.4+8236. ApJ 749:66. https://doi.org/10.1088/0004-637X/749/1/66

Tasse C, van der Tol S, van Zwieten J, van Diepen G, Bhatnagar S (2013) Applying full polarization A-Projection to very wide field of view instruments: An imager for LOFAR. A&A 553:A105. https://doi.org/10.1051/0004-6361/201220882

van Weeren RJ, Williams WL, Hardcastle MJ, Shimwell TW, Rafferty DA, Sabater J, Heald G, Sridhar SS, Dijkema TJ, Brunetti G, Brüggen M, Andrade-Santos F, Ogrean GA, Röttgering HJA, Dawson WA, Forman WR, de Gasperin F, Jones C, Miley GK, Rudnick L, Sarazin CL, Bonafede A, Best PN, Bîrzan L, Cassano R, Chyy KT, Croston JH, Ensslin T, Ferrari C, Hoeft M, Horellou C, Jarvis MJ, Kraft RP, Mevius M, Intema HT, Murray SS, Orrú E, Pizzo R, Simionescu A, Stroe A, van der Tol S, White GJ (2016b) LOFAR facet calibration. ApJS 223:2. https://doi.org/10.3847/0067-0049/223/1/2

Waggett PC, Warner PJ, Baldwin JE (1977) NGC 6251, a very large radio galaxy with an exceptional jet. MNRAS 181:465–474. https://doi.org/10.1093/mnras/181.3.465

Willis AG, Wilson AS, Strom RG (1978) Polarization in the very large radio galaxy NGC 6251 at 610 MHz. A&A 66:L1–L4

Wright EL (2006) A cosmology calculator for the world wide web. PASP 118:1711–1715. https://doi.org/10.1086/510102

Chapter 6
Conclusions

In this thesis I presented low frequency observations of multiple galaxy clusters as well as the giant radio galaxy NGC 6251.

In Chap. 3 I presented GMRT observations of the massive galaxy cluster MACS J2243.3-0935 at 610 MHz. I discovered a new giant radio halo in this cluster with an integrated flux density of 10.0 ± 2.0 mJy, an estimated radio power at 1.4 GHz of $P_{1.4\,\mathrm{GHz}} = 3.2 \pm 0.6 \times 10^{24}$ W Hz^{-1} and a LLS of approximately 0.92 Mpc. I calculated the equipartition magnetic field in the region of the halo for a range of α and K_0 values and find that the equipartition magnetic field is of order 1 μG Assuming a spectral index of $\alpha = 1.4$, the halo in MACS J2243.3-0935 lies on the empirical scaling relations observed for radio halos.

I also detected a potential radio relic candidate to the west of the cluster. The candidate relic has a integrated flux density of 5.2 ± 0.8 mJy, an estimated radio power at 1.4 GHz of $(1.6 \pm 0.3) \times 10^{24}$ W Hz^{-1} and a LLS of 0.68 Mpc. The presence of a radio relic in MACS J2243.3-0935 would make this one of only a handful of clusters that host both a halo and a relic. Due to the position of the relic candidate on the outskirts of the cluster, where a filament meets the cluster, I concluded that the candidate is consistent with an infall relic. I ruled out the possibility of the emission being associated with the WHIM in a filament as the measured flux density and estimated equipartition magnetic field strength are both much larger than expected values for the WHIM. I also excluded foreground galactic emission as an explanation as there is no significant emission in IRIS, SHASSA Hα, WISE or *Planck*.

Future work on this cluster should attempt to properly constrain the spectral index of the radio halo. Observations at frequencies lower than 610 MHZ are important in order to clarify the nature of the candidate radio relic. The low declination of this cluster rules out observations with LOFAR however the newly upgraded GMRT has high sensitivity down to 120 MHz.

MACS J2243.3-0935 was selected for observation based on the highly negative relaxation parameter, indicating a disturbed morphology. Following on from the successful detection of a radio halo in this cluster I observed three more clusters

T. Cantwell, *Low Frequency Radio Observations of Galaxy Clusters and Groups*, Springer Theses, https://doi.org/10.1007/978-3-319-97976-2_6

with highly negative relaxation parameters but covering a wider range of masses. The observations were presented in Chap. 4 of this thesis. I have placed upper limits on the radio power at 610 MHz for three clusters, A07, A1235 and A2055. These limits are below the $P_{610} - L_X$ and rule out bright radio halo in these clusters. I have identified these clusters as potential hosts for USSRH. Observations with LOFAR should be capable of confirming whether or not these clusters host USSRH.

I have begun an observing campaign with the VLA of 10 galaxy clusters at S band in C and D configuration. The purpose of this campaign is to identify new halos and determine the range of Γ where detecting a radio halo is most likely. The criteria used for selecting this sample was that $z > 0.13$ so that the VLA in D-config would be sensitive to emission as large as 1 Mpc. We also required that there be bright central radio galaxies in order to avoid confusion when identifying radio halos.

In Chap. 5 I presented observations of NGC 6251 at 150 MHz with LOFAR HBA. The images presented in this thesis are the highest sensitivity and resolution images of NGC 6251 at these frequencies to date. Analysis of the low frequencies spectral index did not reveal any change in the low frequency spectra when compared with the higher frequency spectral index. NGC 6251 is found to be either at equilibrium or slightly electron dominated, similar to FRII sources. I calculated the ages of the low surface brightness extension of the northern lobe and the backflow of the southern lobe, which are only clearly visible at these low frequencies, to be 205 Myr $< t <$ 368 Myr and 209 Myr $< t <$ 307 Myr respectively. This could indicate that these components are relics of an earlier epoch of activity.

I presented the first detection of polarisation at 150 MHz in NGC 6251, including a weak detection of polarisation in the diffuse emission of the northern lobe. Taking advantage of the high Faraday resolution of LOFAR, I detect Faraday complexity in the knot of NGC 6251 and interpret the weaker component as emission from the lobe located behind the knot. I place an upper limit on the variance in the Faraday depth in the knot of NGC 6251 of $\sigma_{\rm RM} < 5 \times 10^{-3}$ rad^2 m^{-4} and an upper limit on the magnetic field in the group of $B < 0.2\ \mu$G.

Index

T. Cantwell, *Low Frequency Radio Observations of Galaxy Clusters and Groups*,
Springer Theses, https://doi.org/10.1007/978-3-319-97976-2